SpringerBriefs in Electrical and Computer Engineering

More information about this series at http://www.springer.com/series/10059

Dania Marabissi • Romano Fantacci

Cognitive Interference Management in Heterogeneous Networks

 Springer

Dania Marabissi
Università degli Studi di Firenze
Firenze, Firenze, Italy

Romano Fantacci
Università degli Studi di Firenze
Firenze, Firenze, Italy

ISSN 2191-8112   ISSN 2191-8120   (electronic)
SpringerBriefs in Electrical and Computer Engineering
ISBN 978-3-319-20190-0   ISBN 978-3-319-20191-7   (eBook)
DOI 10.1007/978-3-319-20191-7

Library of Congress Control Number: 2015941442

Springer Cham Heidelberg New York Dordrecht London

Printed on acid-free paper

Springer International Publishing AG Switzerland is part of Springer Science+Business Media (www.
springer.com)

Preface

There has been a massive diffusion of wireless devices and applications recently, with a consequent exponential growth of data volume. Future mobile communication systems are challenged to guarantee high-quality and high data rate services to an increasing number of users. Toward this goal several enhancements are considered such as new efficient radio technologies, multiple antenna systems, and carrier aggregation; however, these are important but not decisive to fully match the expected trend. The growing demand of capacity for multimedia traffic triggers new approaches for the wireless mobile network design and deployment. In particular an interesting solution is to increase the access point density, by resorting to the heterogeneous network paradigm. The idea is to improve spectral efficiency by reducing the cell size and by getting the transmitter and receiver closer to each other. Traditional macrocell base stations will be overlaid with small cells characterized by low power and low coverage, thus creating multiple access layers. Differently from the homogeneous macrocellular networks that are usually deployed according to a suitable network planning in order to assure a complete coverage, small cells will be mostly located according to specific traffic needs, forming the so-called hotspots. As consequence, the NetNet paradigm gives rise to new network topologies and interference conditions that need advanced and efficient management methodologies.

This book explains in detail the features of HetNets and outlines the leading key technologies in order to provide an efficient interference management. In particular the book is organized into two parts. Part I, formed by Chaps. 1–2, introduces the main ideas and concepts of HetNets and related interference management techniques. Chapter 1 focuses on possible deployment scenarios and the emerging research challenges. Chapter 2 provides a deeper insight into the possible interference management technologies that can be adopted in a co-channel HetNets deployment, where the macrocell and the small cell share the same resources. Coordinated and uncoordinated approaches are introduced and described. Part II focuses on uncoordinated interference management methodologies and in particular on cognitive approaches. This part is formed by Chaps. 3–4 and constitutes the heart of the book. In particular, an introductory technical overview of cognitive HetNets

v

is given in Chap. 3, where the most important features are introduced based on the generic technologies described in previous chapters. The following Chap. 4 provides a detailed description of possible solutions to mitigate the interlayer interference and optimize the small cell performance by using suitable resource allocation strategies. The solutions are based on the exploitation of the direction of arrival of the signals to separate the users and limit the interference. Finally, looking into the future, Chap. 5 briefly focuses on evolution of HetNets, in order to meet future enhanced requirements.

Since we began working on this book, many people have supported us by providing useful suggestions and valuable contributions. We would like to thank all these people, including our students and valuable colleagues from around the world. Namely, we would like to mention Dr. Giulio Bartoli and Dr. Marco Pucci, our former Ph.D. students, who supported us with valuable technical contributions and discussions during the preparation and writing phases of this book.

A special thanks to Prof. Xuemin (Sherman) Shen from the University of Waterloo, Canada, and also to Mrs Melissa Fearon and Mrs Jennifer Malat who patiently guided the writing of the book.

Firenze, Italy   Dania Marabissi
January, 2015   Romano Fantacci

Acknowledgments

This work has been partially supported by MIUR-FIR FUTURO IN RICERCA Heterogeneous LTE Deployment (HeLD) RBFR13Y0O8_001.

Contents

Part II   Congitive Beamforming Approaches for HetNets

Acronyms

ABS	Almost blank subframe
BS	Base station
CB	Cognitive beamforming
CoMP	Coordinated multipoint
CRE	Cell range extension
CS/CB	Coordinate scheduling/coordinated beamforming
CSI	Channel state information
DL	DownLink
DoA	Direction of arrival
D2D	Device to device
ED	Energy detector
eICIC	enhanced ICIC
eNB	eNodeB
HeNB	Home eNB
HetNet	Heterogeneous network
IA	Interference alignment
ICIC	Intercell interference coordination
ICT	Information and communication technology
ID	Identity
JT	Joint transmission
LTE	Long-term evolution
LTE-A	LTE-advanced
LTE-B	Beyond LTE
MAS	Maximum angle separation
MBG	Maximum beamforming gain
MBS	Macrocell BS
MIMO	Multiple input multiple output
MUE	Macrocell UE
MuSiC	Multiple signal classification
NRT	No real time
OFDM	Orthogonal frequency division multiplexing

OFDMA	Orthogonal frequency division multiple access
PDP	Power delay profiles
PF	Proportional fairness
PRB	Physical resource block
PUE	Primary UE
QoE	Quality of experience
QoS	Quality of service
REB	Range extended bias
RF	Radio frequency
RR	Round robin
RSS	Received signal strength
RT	Real time
RU	Resource unit
SBS	Small cell BS
SCD	Single cycle detectors
SIR	Signal to interference ratio
SINR	Signal to interference plus noise ratio
SNR	Signal to noise ratio
SO	Self-organizing
SUE	Small cell UE
TDD	Time division duplexing
TP	Transmission point
UE	User equipment
UL	UpLink
ZF	Zero forcing
ZFBF	ZF beamforming
2-D	Two dimensions
3-D	Three dimensions
3GPP	Third generation partnership project
4G	Fourth generation
5G	Fifth generation

Part I
Heterogeneous Networks

Chapter 1
Heterogeneous Networks

1.1 Introduction

The volume of mobile data is expected to grow exponentially over the next few years with the massive diffusion of connected devices and the development of increasing numbers of high resource demand applications. Therefore, while the deployment of the fourth generation of wireless mobile broadband systems is accelerating, research is now moving towards fifth generation (5G) networks [2], where the challenge will be to overcome the fundamental limits of the existing cellular networks with the aim of guaranteeing high quality and high data rate services to increasing numbers of users with limited resource availability. 5G wireless communication technologies are expected to achieve 1000 times higher mobile data volumes per unit area, 10–100 times higher numbers of connecting devices and user data rates, 10 times longer device battery lifetimes and a 5-fold reduction in latency.

Therefore, even if there is not yet an official standard for 5G, the main characteristics are emerging. 5G networks will be based on the integration of evolved systems, such as LTE-A and LTE-B, Wi-Fi, and new technologies. Ubiquitous connectivity, QoE satisfaction, environmentally friendly operation, resource usage optimization and cost efficiency represent the main goals for 5G. This means that research activities will have to focus on the technologies and advanced solutions that will allow us to face these challenges. To this end, several important features must be addressed [7, 8]. Among them, HetNet deployment is of special interest, because HetNets are considered to be among the most promising enhancements to the network, and will allow service providers to meet demand for higher transmission capacities.

HetNets represent a novel networking paradigm that is based on the concept of access point densification, and uses a multi-layer architecture (Fig. 1.1) consisting of macrocells overlaid with smaller cells and direct links (i.e., D2D) to serve users with different QoS requirements in a spectrum- and energy-efficient manner [6, 13].

© The Author(s) 2015   3
D. Marabissi, R. Fantacci, *Cognitive Interference Management in Heterogeneous Networks*, SpringerBriefs in Electrical and Computer Engineering,
DOI 10.1007/978-3-319-20191-7_1

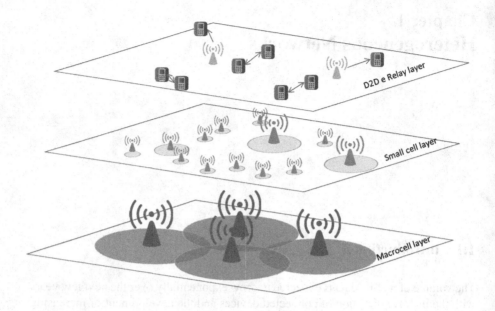

Fig. 1.1 Multilayer heterogeneous networks

Indeed, it is commonly agreed that networks composed solely of macrocells will not be able to satisfy the explosion of mobile data traffic that is currently forecast for 5G.

Small cells are low-power, low-coverage and low-cost base stations that are characterized by different sizes, transmission powers, and backhaul connections. These cells will mainly be developed in areas that require high data transfer peaks or in specific environments where radio coverage must be improved or where reserved access must be guaranteed.

This architectural enhancement, along with the provision of advanced physical communications technology such as high-order spatial multiplexing MIMO communications, will provide higher aggregate capacity for increased numbers of simultaneous users or higher-level spectral and energy efficiencies when compared with those of 4G networks.

Deployment of HetNets is also attractive for operators, because HetNets will allow them to offer extended services, and will also present new market opportunities.

However, despite their potential benefits, the deployment of HetNets must also face the problem of scarcity of available bandwidth. Efficient and flexible solutions that enable operators to use the available spectrum are mandatory to ensure the success of this important technology. Therefore, further spectrum opportunities presented by new frequency bands and application of spectrum sharing policies must also be considered. In particular, two different approaches have been proposed for the deployment of small cells in future wireless networks:

- *Using the same frequency band* as the macrocell;
- using *Separate higher frequency bands* [4, 20].

In LTE, the concept of small cells has been introduced, and refers to a short-range wireless access point that coexists with a macrocell in the same geographical area while sharing the same spectrum. Although it offers some promising advantages, this type of small cell deployment also leads to several practical issues, particularly in terms of inter-layer interference. Indeed, inefficient deployment of the small cells may degrade the performance of the system as a whole. One possible alternative for the evolution of future mobile communication standards is to seek new spectrum resources in higher frequency bands (e.g., beyond 10 GHz) that have not been used in cellular systems. This will lead to deployment of macrocells and small cells in separate frequency bands, which may simplify network operation.

In both deployment scenarios, multi-antenna techniques and spatial domain processing will be vital enabling factors. By considering a spectrum sharing-based deployment, beamforming techniques can be used to reduce both intra-cell and inter-cell interference, while in a separated high frequency deployment, these techniques will be useful in compensating for increased path losses. In particular, the high beamforming gain that is offered by the emerging massive MIMO technology may constitute an essential improvement for small cells. When using the higher frequency bands, common ground can be found between typical millimeter-wave techniques and massive MIMO antenna patterns because of the reduced dimensions of the antenna elements.

1.2 Spectrum Sharing Deployment

In spectrum sharing deployment, the different network layers share the same licensed spectrum. This leads to high spectrum efficiency, but inter-layer interference then forms the main performance bottleneck. Therefore, the benefits of HetNets are strongly dependent on efficient resource management between the high-power macrocells and the low-power small cells. This section focuses on the interference management challenges in this HetNet deployment scenario.

It is argued that existing interference management schemes will not be able to address the interference management problems of HetNets. The coexistence of wide area overlay networks and local BSs leads to specific HetNet topologies, and therefore to new interference environments that cannot be resolved by directly adopting the strategies that are used for traditional macrocell networks. The power imbalance between macrocells and small BSs can significantly limit the capacities of the small cells (*inter-layer interference*), which are also subject to interference from the other small cells (*intra-layer interference*). In addition, to ensure that UE is associated with each of the small cells in sufficient numbers, a mechanism called cell range extension is commonly used [26]. A suitable margin—called the range extension bias—is added to the measured signal strength received from the small cell, so that UE may be assigned to the small cell even when it receives a higher signal from the macrocell, and thus suffers strong interference from the MBSs. Cell association varies also with the user channel access priority, which can be different in each layer.

Another issue that must be taken into account is that low-power BSs can be either user-deployed or operator-deployed. If the low-power nodes are installed directly in an ad hoc manner by end users, and can then be moved, activated, or deactivated at any moment (*uncoordinated deployment*), traditional network planning and optimization become inefficient, because the operator cannot control these nodes. Therefore, new decentralized interference reduction schemes must be introduced that operate independently in each cell, and only on the basis of local information. Low-power nodes must also have cognitive capabilities, i.e., they must be able to monitor the network status and optimize their transmission/reception properties to improve coverage and reduce interference. This type of deployment foresees efficient and opportunistic usage of the spectrum, which is one of the most important elements that must be included in future wireless communication standards. In contrast, when different access node layers are interconnected, it is then possible to achieve efficient coordination among the network layers, and to take dynamic joint decisions about scheduling (*coordinated deployment*).

Finally, there are two different operating modes for small cells that bring different interference scenarios:

- *Closed access mode*—some macrocell users may be within the small cell coverage range, but are not allowed to connect. The SBS may cause high interference with nearby MUEs in DL transmissions, and the MUEs may then interfere with the UL data received at the SBS.
- *Open access mode*—all users are free to connect with the small cell. However, this generates a load balancing problem if each user is associated with the small cell based on only the received signal power; indeed, almost all users would connect to the macrocell, with a consequent reduction in the advantages derived from the heterogeneous development of the network. Therefore, small cell range extension solutions are suggested, which will in turn generate their own interference problems.

A general representation of a heterogeneous deployment scenario using a spectrum sharing approach is shown in Fig. 1.2.

1.2.1 Coordinated Scenario

When different access node layers are interconnected through an interface that is provided for information interchange among the transmission points, such as the X2 interface in LTE-A, it is possible to use coordinated solutions to counteract the interference and handle resource usage.

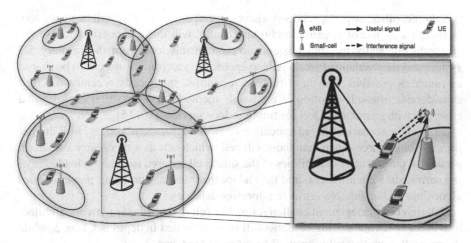

Fig. 1.2 HetNet interference scenario with the spectrum sharing approach [4]

Two main approaches are expected as possible solutions for future broadband wireless networks:

- **Coordinated MultiPoint**: this introduces the possibility of transmitting in a coordinated manner from different network points towards the users that are on the cell edge, and are thus more vulnerable to interference. Different operating modes have been proposed, from simple selection of the best transmission point to joint transmission from different network points or coordinated scheduling and beamforming schemes [25].
- **enhanced InterCell Interference Coordination**: macrocells and small cells use the same frequency channels, but adopt a joint optimal resource allocation scheme with the aim of using orthogonal resources to prevent co-channel interference. Some subframes are thus partially muted by the macrocell to reduce the interference for the most vulnerable SUEs [3, 16]. These special subframes are called almost blank subframes.

CoMP provides high flexibility and improved performance, but requires a large amount of signaling. In contrast, the deployment of eICIC is simpler, but is also less flexible.

1.2.2 Uncoordinated Scenario

Complete coordination among all the cells is often infeasible because of the network delay and signaling overheads, or because the small cells could be user-deployed without any coordination with the network operator and in a closed access mode. Indeed, small cells are likely to be deployed mainly for indoor environments (i.e., home, office, airports, schools) where most of the data traffic will be generated. In these cases, use of cognitive radio approaches will make it possible to use the spectrum efficiently by allowing concurrent transmissions from the primary 5G system and the overlapped small cells to enable the spectrum resources to be shared as much as possible. For these reasons, cognitive small cells recently received considerable research attention as a possible method to provide high capacity and coverage with guaranteed QoS for future indoor services [10, 15].

In such a scenario, the macrocell represents a *primary system* with higher priority for resource usage than the small cell, which acts as a *secondary system*. In particular, the cognitive capabilities of the small cell are used to obtain knowledge of the surrounding environment, and thus adapt the transmission using power control algorithms and/or suitable resource allocation schemes.

Interference management challenges for the future multi-layer network architecture in the spectrum sharing scenario will be investigated in depth in Chap. 2, while the possible cognitive solutions will be discussed in Chap. 3.

1.3 High Frequency Deployment

One of the most widely discussed proposals for systems beyond LTE (i.e., 5G) to overcome inter-cell interference is deployment of small cells in a separate higher frequency band [4]. The limited available spectrum clearly makes dedicated small cell deployment in the current cellular band highly unlikely. However, large bandwidths at higher frequencies ($f_c > 10\,\mathrm{GHz}$) remain unexplored, primarily because of the harsh radio channel propagation conditions. The main challenge is represented by the path loss, which is proportional to a power (usually higher than two) of the carrier frequency; therefore, any signal in a high frequency band would suffer serious attenuation. For example, assuming that the system is moved from 2 to 20 GHz, the received signal would suffer an attenuation penalty of at least 20 dB, which in turn would reduce the system coverage. Therefore, it is of paramount importance that suitable transmission techniques are found for future deployments that improve the quality of the link sufficiently to counteract this problem. In this context, massive MIMO represents a particularly interesting technology. It consists of the use of a very high number of antennas to fully utilize the spatial diversity among the users. One possible application involves formation of very narrow beams to provide a significant gain that can partially counteract the increased path loss. However, one of the major problems with massive MIMO in conventional operating bands is the antenna system dimensions, which may be acceptable for large installations such as a BS, but are completely inappropriate for domestic use. Nevertheless, in a high operating frequency range, the antenna element dimensions become smaller, and thus application of this technology becomes more likely.

As the above discussion shows, high frequency deployment and massive MIMO are closely related. Massive MIMO can be regarded as an enabling technology for high frequency communications, while at the same time, the small antenna size in the high frequency range makes massive MIMO practical. In the next section, the main features of massive MIMO are described.

1.3.1 Massive MIMO

Massive MIMO represents a new research field that spans communication theory, propagation characterization, and the design of associated electronic components. Recently, this topic has attracted major research attention from both academia and industry, because it holds great potential for communication performance improvement [14, 23].

Massive MIMO systems consist of BSs equipped with antenna arrays composed of large numbers (e.g., 100 or more) of small antennas that are plugged together. Ideally, when using a widely separated antenna array, each additional element adds an extra degree of freedom that can be used to introduce highly attractive advantages.

In [19], it is shown that if the number of BS antennas M is much larger than the number of antennas per terminal, K, such that $M/K \gg 1$, then deployment of increasing numbers of antennas is always beneficial; i.e., the effects of fast fading and correlated noise both decrease toward zero, even in low SNR environments. By relying on very simple signal processing, massive MIMO can increase the link capacity while simultaneously improving the radiated energy efficiency [21]. This is achieved by focusing the signal strength in a specific direction to create very narrow radiated beams. It is therefore possible to efficiently transmit independent data flows to different user terminals during the same time-frequency block, thus exploiting the spatial separation of the users (multi-user beamforming). It is easy to see that this strategy can be used symmetrically in the uplink by using the high beam resolution to split (in the angular domain) different signals that are arriving in the same time/frequency slot.

The creation of these very narrow beams also enables inter-symbol interference to be reduced, thus reducing the spectrum inefficiencies of the cyclic prefixes in OFDM symbols. In addition, the total available transmission power is spread among the M available antennas, thus reducing the per-antenna power consumption. In [21] it has been shown that for a BS equipped with a linear detector, massive MIMO can reduce the required amount of per-user transmission power. The amount of power saved is inversely proportional to the size of the antenna array in the case of perfect CSI, and to the square root of the number of antennas for imperfect CSI. This leads to a reduction of the general complexity of the radio-frequency front-end of BSs, and also allows low-cost, low-power components to be used. The use of massive MIMO also leads to a significant reduction in the latency at the air interface, because it dramatically reduces the probability that fading dips will affect the received signals. Finally, robustness with regard to both intentional and unintentional jamming is increased because of the presence of many excess degrees of freedom, which can be used to eliminate any signals coming from jammers.

In conclusion, by implementing small cells equipped with massive MIMO, it is possible to use the available frequency resources efficiently, reduce interference, and ensure improved transmission capabilities. However, some major challenges have still to be addressed.

1.3.1.1 Challenges in Massive MIMO

In theory, massive MIMO systems can achieve all the performance improvements described above. In practice, the design of very large array systems suffers from several implementation issues.

- **Antenna Dimensions.** The realization of massive MIMO systems requires more than 100 small antennas to be placed at the BS front-end. This leads to the implementation of 2D or 3D antenna arrays, where coupling effects among the antenna elements are more obvious, and thus reduces the system capacity. In addition, there is the need for a significant physical space to accommodate the antennas. Deployment in a higher radio frequency spectrum for small cells means that the BS antennas can be designed without the same need for large spaces, thus easing the placement problem.
- **Channel State Information.** Massive MIMO systems must acquire the CSI to operate correctly. Traditionally, in multi-antenna systems pilot symbols are sent by each antenna element in the DL. This is not possible when the number of antenna elements is very high, because the number of pilot symbols would become too high to be practical, especially in mobility scenarios with short channel coherence times. Consequently, massive MIMO systems must operate in the TDD mode, where the channel reciprocity can be used, and the CSI can be estimated at the BS. However, when the number of UEs is high, acquisition of reliable and updated CSI remains challenging, and represents an interesting research topic. In particular, under mobile conditions support of the control signaling and connectivity when operating with highly directive links is not a trivial task.
- **Signal Processing.** When the number of antenna elements increases, the complexity of the algorithms required to determine the suitable precoding vectors or decoding processes also increases. This requires suitable optimization procedures, especially in the case of small cell deployment, because the transmission points have fewer computational resources. Provision of highly accurate narrow beams that can follow individual UEs represents an important issue for investigation, especially when the UEs are mobile. One possible solution is the use of low-complexity algorithms that can be implemented directly in the hardware. The performance loss due to the simplicity of the algorithms would be compensated for by the effects of the high number of antennas.

1.4 Technological Trends in HetNets

In the future, HetNet end-users will have a set of universal connectivity opportunities, and this introduces new and challenging topics for study.

Among these topics, two of the most important are the context awareness and self-organization capabilities of HetNets [12], and concise discussions of the scope and perspectives of these subjects are provided below. These aspects seem to promise new and interesting functionalities to control the local environments where HetNets are deployed by efficiently handling the total incoming traffic load and the mutual interference among the equipment deployed within the HetNet. The desired outcomes should be increased user data rates, better QoS and an important energy-saving advantage.

1.4.1 Context Awareness

The notion and the use of the term *context* has represented an important issue in computer science, and more recently in several emerging domains, including HetNets, performance-oriented applications and autonomic communications [18].

Generally speaking, *context* can be related to any information that can be used to characterize the situation of an entity [24], including its location, identity, time, user preferences and activities. Context is considered to be a general and pervasive concept that may affect the definitions of complex systems at various levels, from the physical device level to the communication level up to the definitions of the users and applications. At each level, contextual information may be crucial, and different processes can be defined that use and adapt to the available contextual knowledge.

In autonomic computing, any device or object, either physical or logical, that is relevant to a specific task is considered to be part of the context [17]. The main challenge in context-aware computing was the definition of a new class of applications that were aware of the context in which they would run. In particular, these applications must be able to adapt to the locations and capabilities of devices, and react to any change over time by modifying either the quality/quantity of results or the way in which the computations are performed [1]. This research trend has led to novel research areas such as context aware HetNets, which can be considered as one current frontier in the well-known field of context-aware computing.

The context information in HetNets is related to knowledge of the specific features of a network entity, and of the status of the entire network (Fig. 1.3). These characteristics can change significantly with the environment. Also, wrong or outdated context information leads to erroneous characterization of the current HetNet status.

In the HetNet environment, context information can be retrieved by the user terminals and the HetNet access points, and this may even involve information exchange between these two types of equipment, or between each of them and a specific database in which the context information is gathered for joint use. Generally, the context information of interest in HetNets, as stated in [1, 17], includes the following:

- Measurements of the RSS or the SINR;
- Direction of the emitting signal source;
- Traffic type and traffic load;
- QoS requirements.

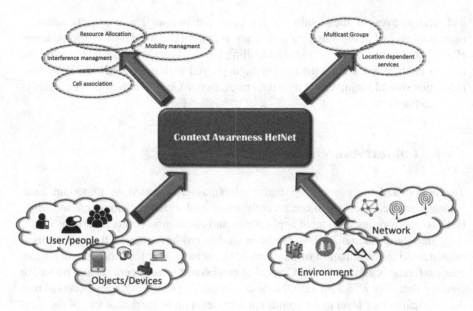

Fig. 1.3 Context Aware HetNets

Two facets of context awareness can be addressed, and it is actually possible to distinguish between *user-oriented* and *self-oriented* context awareness.

User-oriented context awareness means that a context-aware system exposes context-aware services and applications to the users or, in general, to the surrounding environment. However, for HetNets, the context awareness that is achieved on a *self-oriented* basis is of major interest. In particular, this functionality enables HetNets to use context information to allow and drive autonomic behavior, i.e., to allow HetNets to manage themselves. For example, an autonomic HetNet can maintain its performance within the desired limits by dynamically reconfiguring itself using context information that may include the physical location of its clients, their availability, the cost of the communication links, and the criticality of specific activities.

One typical use of context information in HetNets is to drive efficient resource allocation and interference management between the network layers. The metrics used to assess those schemes must be related to both the performance enhancements that they can provide (in terms of both user and system throughput and spectrum efficiency) and the actual complexity that they require in practice.

1.4.2 Self-Organization

Self-organization is an appealing paradigm that enables heterogeneous devices and equipment to perform a task autonomously, even in a collaborative mode. Usually, self-organization refers to a process where some forms of coordination arise from local interactions among the components of the same system or environment.

The self-organization paradigm is a common concept that occurs in a wide class of systems, including physical, chemical, biological, robotic, social and cognitive systems. Common examples include crystallization, the emergence of convection patterns in a liquid that is heated from below, chemical oscillators, swarming in groups of animals, and the way that neural networks learn to recognize complex patterns [9]. Self-organization now represents a rather attractive capability in HetNets that would enable reduced human intervention, and thus noticeably reduce deployment and management costs [22]. The basic idea of SO-HetNets involves consideration of network planning, configuration and optimization as an overall process that must be handled by minimizing human interactions. A SO-HetNet must have the basic capabilities of any SO system, as listed below:

- *Self-configuration*—this refers to methodologies used to identify or modify a suitable configuration, depending on the service needs or environmental conditions. These methodologies use observation of the system environment, including the QoS requirements, interference, and the interactions with all HetNet components. Self-configuration in HetNets usually refers to a process that enables automatic configuration of any newly deployed HetNet device to set basic parameters for efficient operation.
- *Self-management*—this refers to skills for efficient handling and preservation of HetNet devices. Generally speaking, this is not a technique, but is actually a specific feature of a SO-HetNet, which means that each HetNet component must be able to control and preserve itself, depending to the actual service requirements and the environmental conditions.
- *Self-healing*—this refers to techniques used for efficient reaction to HetNet device faults. This is quite an important feature because it allows the HetNet to tolerate faults. Self-healing techniques must allow the detection, classification and repair of failures in an autonomic manner and without human intervention.
- *Self-optimization*—this refers to the capability of the HetNet to optimize its operational parameters with respect to specific goals.

To date, HetNets still do not offer SO features, mainly because the architectures of SO networks, as specified by the 3GPP, are designed primarily for the homogeneous network case. However, it is widely accepted that SO capabilities are essential to utilizing the advantages of HetNets. In particular, SO-HetNets will provide many important benefits in terms of improved QoS, energy savings, and resilience, and will allow operators to overcome several serious problems related to network planning, such as inter-cell interference and resource assignment.

1.5   HetNet Applications

HetNet technology is of interest for a wide class of applications, including residential, enterprise and urban applications, to allow networks that are always on and are available anywhere at any time.

- *Residential.* In the residential application case, the small cells will mainly be user-deployed low-power access points, and their main goal will be to offer wireless broadband services to residential users, which are usually provided with reserved access. Indeed, it is expected that most data traffic in the future will originate indoors, and HetNets will offer an efficient and cost-effective solution to reduce the amount of traffic that is carried by the macrocellular network by offloading the traffic onto the small cells, particularly for heavy data users. Simultaneously, closed-access small cells can assure the QoS and guarantee access for the residential users. Self-configuration and optimization capabilities will be of paramount importance in this application context for HetNets to enable a plug-and-play usage mode (i.e., easy installation). In addition, the small cell BSs must be compact.
- *Enterprises.* HetNets for enterprise applications will generally be based on a dense deployment of small cells, and will be used in shopping centers, hotels, hospitals, stadiums and similar locations. HetNets are thus tailored based on specific application requirements. As an example, HetNets can enable real-time location services that allow employees to use their smartphones as IDs to access offices, or can provide dedicated services to both terminal and transit passengers in an airport. In this environment, the small cells will be either user-deployed or operator-deployed.
- *Urban.* In urban applications, HetNet deployment is a promising technology that not only offloads traffic from the macrocell, but also helps to solve radio coverage problems. In this case, small cells are deployed in specific areas where the signal received from the macrocell is low or where high data loads are expected (i.e., hot spots), such as theme parks, sporting venues and subways. In this case, the HetNets must be scalable to allow extension of the network coverage and capacity, and to enable the integration and interoperability of multi-technology radio networks.

  In this environment, HetNets appear to be of special interest for the emerging paradigm of the smart city, as a means to increase the use of intelligent ICT solutions in an urban ecosystem. The "smart" concept is based on functional integration of software systems, network infrastructures, heterogeneous user devices and collaboration technologies. In this scenario, small cells can provide specific services and applications to enable smart infrastructures, such as:

  - City management: improves the quality of services offered by reducing costs, while increasing transparency and citizen participation. Small cells can create specific hot-spots for access to services, and also increase in-home connectivity.

– Healthcare: allows more accurate and rapid diagnosis and emergency response. The use of small cells can allow the retrieval of micro location information, thus enabling more accurate localization of the emergency, even in an indoor scenario.

– Transportation: improves services, because traffic management and multimedia information delivery combined with reduced costs encourages the use of public transportation. Small cells can improve these services in congested areas or in coverage holes (e.g., tunnels).

– Environment: improves environmental conditions by limiting sources of pollution, increasing green areas, and providing widespread health services. Small cells represent a viable solution to reduce power emissions, and work towards the paradigm of green networks.

– Public safety: improves capabilities for monitoring of critical infrastructures and forecasting of accidental or intentional (e.g., due to terrorist attack) disasters [11]. This allows the use of real-time information to provide rapid response to emergencies and threats. Small cells can improve both surveillance and rescue operations, especially in specific areas such as critical infrastructures.

Finally, the ubiquity of smartphones and the improved availability of broadband connections owing to the presence of HetNets will make citizens a primary source of real-time information, even in areas where classical monitoring techniques are either unavailable or not accessible. In particular, social networks represent an innovative tool for information gathering and broadcasting, and allow the gathering of information from and dissemination of information to the general public, even in a public safety framework. A practical usage case related to the combined use of advanced communication systems, sensor networks and social networks for a smart public safety application is discussed in detail in [5].

References

1. Andreas N.D., Kliks A. and Holland O. (2014) Context-Aware Radio Resource Management in HetNets. Proc. of IEEE Wireless Communications and Networking Conference (WCNC)
2. Baker M. (2012) From LTE-advanced to the future. IEEE Communications Magazine 50(2):116–120
3. Barbieri A., Damnjanovic A., Tingfang Ji, Montojo J., Yongbin Wei, Malladi D., Osok Song and Horn G. (2012) LTE Femtocells: System Design and Performance Analysis. IEEE Journal on Selected Areas in Communications 30(3): 586–594
4. Bartoli G., Fantacci R., Letaief K., Marabissi D., Privitera N., Pucci M. and Jun Zhang (2014) Beamforming for small cell deployment in LTE-advanced and beyond. IEEE Wireless Communications 21(2): 50–56
5. Bartoli G., Fantacci R., Gei F., Marabissi D. and Micciullo L. (2013) A novel emergency management platform for smart public safety. International Journal of Communication Systems. Article first published online
6. Bhushan N., Junyi Li, Malladi D., Gilmore R., Brenner D., Damnjanovic A., Sukhavasi R., Patel C. and Geirhofer S. (2014) Network densification: the dominant theme for wireless evolution into 5G. IEEE Communications Magazine 52(2):82–89

7. Boccardi F., Heath R.W., Lozano A., Marzetta T.L. and Popovski P. (2014) Five disruptive technology directions for 5G. IEEE Communications Magazine 52(2): 74–80
8. Demestichas P., Georgakopoulos A., Karvounas D., Tsagkaris K., Stavroulaki V., Jianmin Lu, Chunshan Xiong and Jing Yao (2013) 5G on the Horizon: Key Challenges for the Radio-Access Network. IEEE Vehicular Technology Magazine 8(3):47–53
9. Dresler F. (2007) Self-Organization in Sensor and Actor Networks. Wiley Series in Communications Networking and Distributed Systems, J. Wiley
10. ElSawy H., Hossain E. and Dong In Kim (2013) HetNets with cognitive small cells: user offloading and distributed channel access techniques. IEEE Communications Magazine 51(6): 28–36
11. Fantacci R., Marabissi D. and Tarchi D. (2010) A novel communication infrastructure for emergency management: the In.Sy.Eme. vision. Wireless Communications and Mobile Computing, John Wiley & Sons, Ltd. 10(12):1672–1681
12. Fratu Octavian, Vulpe Alexandru, Craciunescu Razvan and Halunga Simona (2014) Small Cells in Cellular Networks: Challenges of Future HetNets. Wireless Personal Communications 78(3): 1613–1627
13. Hu R.Q. and Yi Qian (2014) An energy efficient and spectrum efficient wireless heterogeneous network framework for 5G systems. IEEE Communications Magazine 52(5):94–101
14. Larsson E., Edfors O., Tufvesson F. and Marzetta T. (2014) Massive MIMO for next generation wireless systems. IEEE Communications Magazine 52(2):186–195
15. Lien Shao-Yu, Kwang-Cheng Chen, Ying-Chang Liang and Yonghua Lin (2014) Cognitive radio resource management for future cellular networks. IEEE Wireless Communications 21(1):70–79
16. Lopez-Perez D., Guvenc I., De la Roche G., Kountouris M., Quek T.Q.S. and Jie Zhang (2011) Enhanced intercell interference coordination challenges in heterogeneous networks. IEEE Wireless Communications 18(3):22–30
17. Makris P., Nomikos N., Skoutas D., Vouyioukas D., Skianis C., Zhang J. and Verikoukis C. (2013) A context-aware framework for the efficient integration of femtocells in IP and cellular infrastructures. Eurasip J. Wireless Communications and Networking
18. Makris P., Skoutas D. and Skianis C. (2013) A survey on context-aware mobile and wireless networking: On networking and computing environments integration. IEEE Communications Surveys Tutorials 15(1): 362–386
19. Marzetta T.L. (2010) Non cooperative Cellular Wireless with Unlimited Numbers of Base Station Antennas. IEEE Transactions on Wireless Communications 9(11): 3590–3600
20. Nakamura T., Nagata S., Benjebbour A., Kishiyama Y., Tang Hai, Shen Xiaodong, Yang Ning and Li Nan (2013) Trends in small cell enhancements in LTE advanced. IEEE Communications Magazine 51(2):98–105
21. Ngo Hien Quoc, Larsson E.G., Marzetta T.L. (2013) Energy and Spectral Efficiency of Very Large Multiuser MIMO Systems. IEEE Transactions on Communications 61(4):1436–1449
22. Peng Mugen, Liang Dong, Wei Yao, Li Jian and Chen Hsiao-Hwa (2013) Self-Configuration and Self-Optimization in LTE-Advanced Heterogeneous Networks. IEEE Communication Magazine: 36–45
23. Rusek F., Persson D., Buon Kiong Lau, Larsson E.G., Marzetta T.L., Edfors O. and Tufvesson F. (2013) Scaling Up MIMO: Opportunities and Challenges with Very Large Arrays. IEEE Signal Processing Magazine 30(1): 40–60
24. Schilit B. and Theimer M. (1994) Disseminating active map information to mobile hosts. IEEE Network 8(5): 22–32
25. Shaohui Sun, Qiubin Gao, Ying Peng, Yingmin Wang and Lingyang Song (2013) Interference management through CoMP in 3GPP LTE-advanced networks. IEEE Wireless Communication 20(1):59–66
26. Qiaoyang Ye, Beiyu Rong, Yudong Chen, Al-Shalash M., Caramanis C. and Andrews, J.G. (2013) User Association for Load Balancing in Heterogeneous Cellular Networks. IEEE Transactions on Wireless Communications 12(6):2706–2716

Chapter 2
Interference Management in HetNets

2.1 Introduction

The coexistence of heterogeneous cells that have different characteristics (including power, coverage, backhaul links, and density) leads to specific topologies and new interference scenarios that cannot be solved by direct application of the solutions that have previously been used for traditional single layer networks [18, 40]. In addition, the access methods that are used in small cells have an enormous impact on the overall interference. Indeed, the interference is more severe in closed access mode when compared with the open access alternative, because any UE may be assigned to a macrocell, even if it receives a higher signal from the closed access small cell, which then becomes a strong interference source.

In general, two types of interference can arise in a multi-layer network architecture:

- *Intra-layer interference*: this type of interference occurs among elements located on the same layer. In particular, while intra-macrocell interference can be limited by accurate cell layout planning, small cells can also be randomly placed, which may result in cells being placed very close to each other, thus creating more interference problems. Small cell UEs can cause UL interference with the neighboring small cell BSs. In turn, these small cell BSs can cause DL interference that affects UEs that belong to neighboring small cells.
- *Inter-layer interference*: this type of interference occurs among the elements of different layers, and in particular among macrocells and small cells. A small cell BS can produce interference with the DL of a nearby macrocell UE, while the macrocell UE can produce interference with the UL of a nearby small cell BS.

Interference arises whenever the same radio resource is used simultaneously by different cells. Therefore, system performance is determined by the probability of

© The Author(s) 2015
D. Marabissi, R. Fantacci, *Cognitive Interference Management in Heterogeneous Networks*, SpringerBriefs in Electrical and Computer Engineering,
DOI 10.1007/978-3-319-20191-7_2

collision and how such a collision will affect the SINR. In general, interference management techniques aim to reduce the collision probability and mitigate SINR degradation using appropriate resource allocation strategies [19].

Various solutions can be considered, which are usually related to a combination of the specific scenario and the small cell capabilities [36]. In *static resource allocation schemes*, the resources are split among the cells during the radio planning process. These schemes represent less complex but less efficient strategies. Conversely, in *dynamic resource allocation schemes*, resources are dynamically assigned to the cells, depending on the instantaneous traffic load. The control of these decisions can be either centralized or distributed. In the first case, the resources are allocated using either local information from users or information from other BSs, while in the second case, only the local information is used. Hybrid approaches that achieve the desired trade-off between complexity and efficiency are possible, and take the backhaul capacity into account. Indeed, the backhaul is an important issue that drives the selection of possible interference strategies. When high amounts of signaling can be exchanged among the cells, it is then possible to adopt efficient and fully coordinated centralized schemes; however, when the backhaul connection among the cells is either limited or unavailable, only partial or even zero coordination is allowed.

Interference management in HetNets is also strongly dependent on optimization of the cell association policy. This allows operators to improve the system performance in terms of increased throughput, interference reduction and macrocell traffic offloading simultaneously.

Finally, most of the interference management schemes that are used in HetNets assume user classification based on their average SINR as a basic concept. Therefore, different interference management strategies must be adopted to reduce the performance degradation of UEs that belong to different classes (which are also known as cell regions). In particular, UEs that are located close to the cell boundary suffer from low useful signal levels and high interference (i.e., low SINR) values, and as a result are among the most critical UEs in terms of the QoS achieved.

2.2 Cell Association

The classical cell association strategy is based on the cell SINR values. A UE selects the cell with the highest measured SINR value based on specific DL reference signals to act as a serving cell. This criterion maximizes the cell throughput in a single layer network, but also leads to a huge traffic load imbalance in HetNets. Indeed, small cells are characterized by low transmission power and thus by reduced coverage that is also limited by interference from the macrocell BS, which has significantly higher transmission power. Therefore, when using the traditional approach, only UEs that lie in close proximity would select the small cell as a serving cell, which makes this approach unsuitable for HetNets. As a consequence

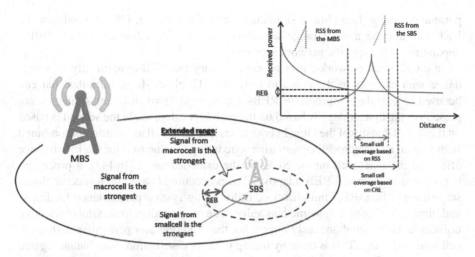

Fig. 2.1 Cell range extension

of this approach, UEs may connect to distant high-power macrocells rather than to any nearby small cell, thus leading to inefficient load distribution.

To reduce the macrocell load and maximize cell splitting, the small cell coverage area can be extended using the *cell range extension* strategy. This method is based on the use of a bias (i.e., REB) that comes from the small cell and is a positive value that is added to the measured signal power received by the UEs (Fig. 2.1).

As shown in Fig. 2.1 the UEs in the extended range experience low SINR conditions because they are no longer connected to the strongest cell, which instead could cause severe interference for these UEs. For this reason, CRE cell association is usually combined with suitable resource partitioning policies, as detailed in Sect. 2.3.1.

CRE is a natural enabler of offloading, as shown in [48], where the cell association problem is studied in detail. In particular, the authors have shown that CRE is a simple approach that allows users to achieve results that are very close to those that can be obtained using a load-aware utility maximization function, provided the REB values are chosen carefully. Indeed, despite the fact that many of the works on CRE that are available in the literature assume fixed REB values [41, 45], optimal selection of this parameter strongly affects HetNet performance, in that

- Low REB values lead to a small number of UEs being connected to the small cell, with consequent load imbalance and inefficient use of the small cell resources;
- High REB values lead to a high degree of extension of the small cell coverage, which could then be congested by UEs with poor SINR values.

In [16], a semi-analytical evaluation of the network capacity and the fairness achieved using a combination of CRE and a resource partitioning method is performed for different REB values and illustrates the need for suitable network parameter tuning. In addition, simulation results reported in [38] showed that the REB can assume a wide range of values (ranging from few dB up to 20 dB), depending on the specific network scenario.

In coordinated networks, it is possible to vary the REB dynamically in accordance with changes in the scenario [39]. In [31], closed-form formulas that can be used to calculate appropriate REBs are derived. Two different strategies are presented; the first strategy is based on the path loss value, while the second is based on the spatial density of the UEs. In both cases, the CRE cell association is combined with a small cell-macrocell cooperative scheduling scheme to limit the interference effects of the macrocell on the SUEs in the extended range. In [43], a procedure is proposed to make the REB adaptive. The procedure is based on a specific three-step strategy: first, REB initialization, followed by system performance feedback, and, finally, setting of a dynamic bias value. The scheme aims to maximize resource utilization while simultaneously improving the cell edge user performance through cell load balancing. This is done by taking the user distribution, load balancing and fair service quality among all users into account.

The benefits of suitable cell association strategies for HetNets can be improved if they are combined and optimized in conjunction with the resource allocation procedures. The CRE association simply forces UEs to select low power nodes by adding a fixed bias to the received signal power, but it does not take the probability that an idle resource in the selected small cell is assigned to the UE into account. This probability depends on both the resource allocation strategy and the number of UEs in the small cell. For example, if the small cell uses an RR allocation scheme and has a large number of UEs, it may keep the off-loaded users in a starvation state. This problem can be resolved by reducing the REB value. Alternatively, a different cell association policy that is resource-aware can be adopted. For example, a UE can be associated with a cell that has the maximum probability for allocation of a channel to that UE, independently of its REB value.

2.3 Coordinated and Cooperative Interference Management Approaches

Coordination and cooperation among the different layers are promising approaches with which to address the interference problems in HetNets. These methods are made possible when both small cells and macrocells are deployed by the same operator and a backhaul link is available. In this case, the different layers are interconnected through specific interfaces that are devoted to information exchange among the network elements.

With cooperative approaches in particular, a UE is allowed to be connected simultaneously to more than one cell, and this allows the UE to select the best connection or to make combined use of the multiple connections. This approach requires tight integration of the cells into the network, and fast, low latency backhaul links to exchange the signaling and data information. The cooperative strategy that is most widely considered in LTE-A networks is the CoMP strategy (Sect. 2.3.2).

Coordinated approaches instead allow the constraints on the backhaul links to be relaxed, because in this case, each cell serves only its own UEs and therefore only signaling information must be exchanged among the cells. The best known coordinated strategy is the eICIC strategy (Sect. 2.3.1).

2.3.1   Enhanced ICIC

eICIC represents an evolution of previous static ICIC schemes based on frequency partitioning, as outlined in [17], to provide greater dynamism and flexibility. The eICIC concept relies on time-division-based resource splitting among the macro-cells and small cells [30]. In particular, the macrocell BS can reduce interference with the SUEs by using the ABSs according to a preconfigured pattern. In the ABSs, the macrocell only transmits cell reference symbols and some control messages, which results in low-power interference for the UEs in the small cell served by these resources. Consequently, a small cell can extend its coverage area during ABSs because these subframes are no longer dominated by the macrocell interference. During ABSs, the small cell can communicate with the most vulnerable UEs (which are usually the UEs in the CRE area) with reduced interference levels. Other UEs are scheduled during the non-ABSs.

The eICIC concept is represented schematically in Fig. 2.2.

The ABS muting pattern is characterized by the positioning of the ABSs in the frame and by the ratio of the ABSs to the total number of subframes, which is called the *ABS muting ratio*. To exploit the ABSs appropriately, the small cell must know the muting ratio used by the macrocell BS and schedule the UEs accordingly. The ABS pattern is periodic and is not specified by a standard. Usually, the pattern is left to the operator for implementation, and it can be changed dynamically to find the configuration that maximizes overall system performance and guarantees the target QoS for the UEs. The value of the ABS muting ratio affects the average data rates of the offloaded UEs and depends on the number of users that have been offloaded. ABS pattern selection should also take either the benefits of the offloaded UEs or the macrocell throughput reduction due to loss of resources into account.

ABS pattern selection is considered in several papers, e.g., [3, 10, 25, 29, 44], where various methods are proposed to dynamically determine the most suitable ABS ratio value, depending on the specific goals and the assumptions used. In addition, because of their strong correlation, the REB and the ABS pattern can also be jointly optimized, as outlined in [6, 11].

The definition of the ABS can be extended by also including subframes in which the macrocell BS is not muted, but where data is transmitted with reduced power to provide greater system flexibility and to permit better balancing between small cell gain and macrocell loss [3].

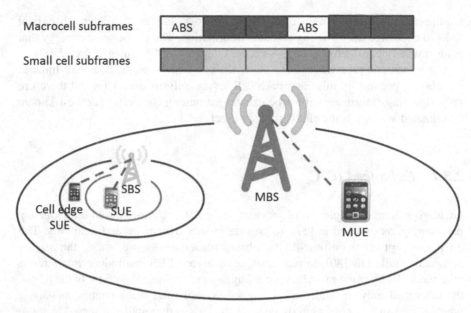

Fig. 2.2 eICIC concept

2.3.2 Coordinated Multi-Point

CoMP approaches can be considered to be an evolution of the eICIC schemes. In particular, the CoMP evolution brings greater flexibility, ranging from a static or semi-static approach to dynamic joint decisions about scheduling. CoMP is adopted to increase spectral efficiency, particularly at the cell edge.

Two CoMP approaches are envisaged that differ mainly in terms of the required network signaling overhead; these approaches are:

- *Joint transmission*;
- *Coordinated scheduling/coordinated beamforming*.

With JT, data streams are transmitted simultaneously from the cooperating TPs towards the same user.

With CS/CB, data that is addressed to a specific UE is transmitted by a single TP, while other TPs use the same radio resources to transmit to spatially separated UEs by using beamforming techniques. JT can introduce a high throughput gain for the UEs that are served, but JT also requires a high backhaul capacity because of the need to share CSI and data among the TPs. Conversely, CS/CB only needs to share the CSI and scheduling information, and is therefore more suitable when the backhaul link is not ideal, as is the case in many practical networks. The two possible CoMP approaches are depicted in Figs. 2.3 and 2.4.

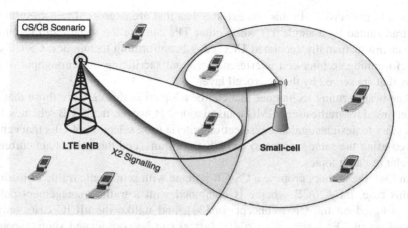

Fig. 2.3 CS/CB CoMP approach [2]

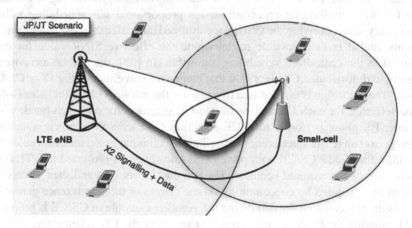

Fig. 2.4 JT CoMP approach [2]

Resource allocation can be managed with:

- *Hierarchical schemes*, where a BS set elects a master (which is usually a BS or another network entity) to be in charge of coordinating certain operations and decision making on behalf of the group;
- *Horizontal schemes*, where the same algorithm runs at each CoMP node and thus returns the same resource allocation pattern.

2.3.2.1 CS/CB CoMP

The CS/CB CoMP scheme is based on sharing of the information provided by each UE to its TPs, along with other information exchanged among the TPs, to make a joint decision. In this context, the data that are addressed to a specific UE are transmitted by a single TP, while other TPs can use the same radio resources to transmit to spatially separated UEs using beamforming techniques. CS/CB can therefore mitigate inter-cell interference without sacrificing the throughput of the users that are served by the macrocell layer.

The beamforming techniques that can be adopted in this case are those that are widely used for multi-user MIMO transmissions. However, in CS/CB schemes, it is necessary to design suitable UE selection criteria (i.e., selection of UEs that can be served using the same resources) and CS algorithms, and these areas are currently popular research topics.

In [34], the authors propose a CS/CB scheme with semi-static traffic offloading. In this case, the CS/CB scheme is combined with a traffic management policy that is based on the CRE concept. In [34], and unlike the eICIC case, only a limited set of TPs (called the *CoMP cluster*) can be coordinated simultaneously to reduce the interference. The CS/CB scheme works by following a classical zero forcing beamforming procedure and a proportional fair scheduling policy. Specifically, after receiving the CSI, the centralized unit calculates the beamforming weights, and it is then possible to determine the effective SINR value for each UE used in the centralized scheduling algorithm. In [23], the authors propose an adaptive and distributed CS algorithm that can operate in either of the JT or CS/CB modes. The algorithm is divided into two phases: the intra-BS phase includes CoMP mode selection for each UE and candidate UE selection for each sub-band, while the inter-BS phase performs final UE selection for each sub-band via a suitable message exchange. An interference cancellation technique based on a user selection algorithm for CoMP CS/CB with partial CSI knowledge is proposed in [22]. Each TP determines the required beamforming weights for its UEs and then selects the users to be scheduled by concentrating on the effects of the interference generated towards users located on the cell boundary. Another example of CS/CB is proposed in [5], where a criterion is first introduced to select the CB scheme that is most suitable for a randomly chosen UE pair that share the same resources. Then, a CS method that allows the most suitable UE pairs to be created is proposed. The goal of the proposed procedure is to maximize cell throughput while simultaneously guaranteeing fairness among the cells. Either the CB scheme or the CS scheme must be based on knowledge of the DoA of the UE signals.

2.3.2.2 JT CoMP

The JT scheme can be used when higher signaling overheads are supported by the network and more sophisticated cooperation processes are allowed. In the JT scheme, data transmission towards a UE is performed using multiple TPs simultaneously. This means that the user data must be available at all TPs, and thus demands high backhaul link capacity. The UE can combine multiple copies of a signal to improve the received signal quality. The JT can be implemented using either coherent or non-coherent approaches, depending on whether the signals are summed coherently (i.e., by taking the signal phase into account) or not. In general, coherent combining produces the best performance but is more sensitive to CSI errors. Non-coherent JT cannot nullify the interference because of a lack of phase information across the TPs.

JT-CoMP is considered to be a very promising solution. In the literature, the studies of JT-CoMP in HetNets deal with resource scheduling, with the goal of maximizing throughput for the cell edge users [15, 26], and with achieving efficient use of power to produce solutions that are compliant with green communication paradigms [14, 47].

The joint use of CoMP transmission with CRE strategies is analyzed in [9]. Specifically, the authors consider three different CRE strategies and compare them in terms of load balancing. These strategies are based on a classical positive bias that is added to the RSS (see Sect. 2.2), on the path loss, and finally on a hot-spot area. CoMP is used to counteract the increase in inter-cell interference for the SUEs in the extended range.

Many works in the literature deal with the main challenge related to CoMP implementation, which is the increased signaling overhead because of inter-BS communications and the consequent backhaul problems. In practice, only a limited number of BSs (i.e., the *CoMP cluster*) could cooperate, jointly transmitting the data towards the UE. TPs within the same cluster then coordinate themselves to serve the users in this cluster, and can also perform some forms of inter-cluster interference management. Cluster detection in a realistic scenario has also been a subject of major study. The proposed solutions are based on either static or dynamic clustering. As an example, in [33], two static clustering algorithms are presented and evaluated in both an ideal hexagonal cell layout and a real-world deployment. The first scheme is intended to maximize the mean SINR value, while the second is intended to maximize the number of locations at which a certain target SINR value can be achieved. The authors consider a cluster that is made from a fixed number of cells, M, and two contrasting alternatives: non-overlapping and overlapping clusters. It is shown that the proposed static clustering method has a performance that is close to that of a UE-specific clustering method, which means that each UE is served by the M received cells with higher SINR. In contrast, a novel dynamic greedy algorithm for CoMP cluster formation is presented in [37]. The goal of the clustering algorithm is sum-rate maximization of the UEs that were selected via an RR scheduling algorithm. The paper shows that dynamic clustering schemes outperform static clustering schemes because of their ability to adapt to both traffic demand and UE locations. In [50], the authors address the problem of joint optimization of both the BS assignment for each UE and the beamformer design, subject to per-BS power constraints and given the QoS requirements. Their aim is to minimize backhaul user data transfer from the data center to the BSs that belong to the cluster. Because of the complexity of the solution to the optimal problem, the authors propose a sub-optimal heuristic. The intra-BS information exchange can also be reduced, limiting the quantity of feedback that is sent by the UEs. As an example, in [24], the UE only feeds back the CSI for a set of BSs that have signal strengths exceeding a specific threshold.

2.3.2.3 eICIC vs CoMP

As stated previously, the joint use of eICIC and CRE cell association represents a simple solution that requires only limited coordination among the BSs and, consequently, low signaling overheads. However, optimization of the eICIC parameters is not a simple task. The task depends on multiple factors and a general optimum selection criterion is hard to define. In addition, the ABS pattern is fixed by the macrocell throughout its coverage area, while the REB can be varied independently among the cells based on their conditions (including traffic load, number of UEs, and propagation links). This could lead to a disproportionate relationship between the muting ratio and the bias. Finally, the eICIC is considered as a semi-static coordination scheme that is unable to respond to instantaneous changes in network topology. In contrast, the CoMP technique offers high flexibility and can follow rapid topology and traffic load changes. It is also very effective in increasing the performance levels of the cell-edge UEs. However, CoMP is strongly limited by both the CSI feedback overhead and the backhaul requirements, and this suggests either adaptive or ad hoc use of CoMP.

2.4 Cognitive Interference Management Approaches

In both cooperative and coordinated approaches, the requirement for information exchange between the nodes introduces problems with regard to the backhaul link capacity and, consequently, the high cost of network deployment. One of the most attractive features of small cells is their ad hoc installation, which allows mobile operators to save on backhaul costs because the traffic of a small cell can be carried by subscriber broadband communications links. However, when small cells are deployed ad-hoc, there is either limited or zero coordination with the macrocells, and thus it is difficult to manage small cells from a centralized controller based on cell planning with resource partitioning between the two network layers. In this case, it is preferred that intelligence is induced in the small cell to enable it to counteract inter-cell interference without access to any centralized controller. In addition, different small cells, depending on their location and the surrounding environment, would each face different interference scenarios; therefore, it is important that small cells have self-organization capabilities (see Sect. 1.4.2). The interference can be managed using a cognitive radio approach that allows concurrent transmissions that arise from macrocells and small cells to share the spectrum resources [27]. This solution requires the small cell to have cognitive capabilities that allow it to acquire environmental awareness and to autonomously identify the presence of other cells via a periodic sensing process that is essential for identification of the unoccupied portions of the spectrum. Knowledge of the surrounding environment is used to adapt both transmission and reception to limit mutual interference. The cognitive capabilities thus require a combination of opportunistic and dynamic resource allocation schemes with advanced signal processing at the receiver end [21]. The general concepts of cognitive HetNets are presented in this section, while a more detailed discussion is provided in the next part of the book.

2.4.1 Cognitive Radio Principles

One specific feature of future 5G networks is that the spectrum usage efficiency will be increased by allowing sharing by different systems. CR methodologies seem to offer an attractive route towards this goal.

In a classical CR approach, it is necessary to distinguish between the primary and secondary systems. Specifically, the secondary system must be able to use the same resources that are allocated to the primary system using suitable strategies to ensure that the secondary system does not affect the primary systems performance.

CR methodologies are well known to provide an interesting research area, but, despite this, there has been a lack of widespread applications of CR techniques to date. An attempt to apply these techniques will start shortly based on the use of white spaces in television signals. However, the use of CR schemes in future 5G communication networks is mandatory for effective spectrum sharing among different systems and the provision of new services.

In general, CR technology is based on two main characteristics: cognitive capability and reconfigurability [13]. Cognitive capability refers to the ability to acquire context awareness. Reconfigurability enables a secondary network to be dynamically adapted to the available radio environment. More specifically, CR can be designed to adapt its transmission using power control algorithms and/or suitable resource allocation schemes.

Two different cognitive approaches can be identified:

- *Opportunistic approach*;
- *Underlay approach*.

When following an *opportunistic approach*, the secondary system can use portions of the radio resources that are unused by the primary system in a given time and space. The secondary cognitive system therefore uses radio resources dynamically on an opportunistic and non-interfering basis by exploiting frequency holes. An alternative cognitive approach is presented by the *underlay approach*. In this method, the secondary system can share the channel simultaneously with the primary system by adopting constraints on its power emissions to reduce (or avoid) mutual interference. In both cases, the secondary system must sense the radio channel to determine the resources that are idle among the available choices or the resources that can be used while introducing only a limited amount of interference. In addition, a cognitive system must adapt to changes in the surrounding environment. This means that both sensing and reconfiguration must be performed periodically.

Two different levels of spectrum sharing can be addressed; specifically, the transmission opportunities for the secondary system are:

1. The portions of the available spectrum (i.e., sub-bands) that are not used by the primary system during a certain time period;
2. The smallest resource units that are idle on a given frame.

The main difference between the two levels is the time scale. In the first case, it is reasonable to assume that the transmission opportunities will last for several frames, while in the second case, the resource assignment will change frame by frame. For this reason, the two cases are referred to as *Not Real Time* and *Real Time* cognitive approaches in the following.

2.4.1.1 NRT Cognitive Approaches

In future 5G networks, the need for wide frequency bands and the lack of availability of large portions of the free spectrum will lead to the aggregation of even non-contiguous bandwidth fragments: flexible and scalable frequency usage will become possible by adopting dynamic sub-band aggregation processes to satisfy the demand for requested capacity. Therefore, each BS will change the number of sub-bands used adaptively, depending on both the traffic load and the interference level from/towards neighbor cells.

In LTE-A systems in particular, this important feature is called *carrier aggregation* and allows aggregation of a maximum of five component carriers (indicated by sub-bands B_i in Fig. 2.5) over a bandwidth of up to 20 MHz to attain total transmission bandwidths of up to 100 MHz. Because none of the available service providers owns a continuous spectrum of 100 MHz, three different carrier aggregation modes exist: intra-band contiguous and non-contiguous aggregation, and inter-band aggregation, as shown in Fig. 2.5.

In a two-layer cognitive network, the sub-bands can be used opportunistically by the cells; if the macrocell (i.e., the *primary system*) uses only a sub-set of the available frequency sub-bands, then the remaining sub-bands can be used by the small cells (i.e., the *secondary system*), thus allowing coexistence between the two layers without interference. This scenario is depicted in Fig. 2.6.

In [28, 35, 46], several examples of opportunistic usage of the available frequency sub-bands among different cells are presented. These schemes work in a coordinated manner based on exchange of information among the network layers. Conversely, in a cognitive scenario, the small cell must be able to detect the use of a given sub-band by performing appropriate spectrum sensing. The sub-bands in which no activity is detected from nearby macrocells are used to communicate via the small cell. It is reasonable to assume that the primary system sub-band aggregation does not change frame by frame, but the process is performed periodically when the network load changes significantly or when new cells are activated in the same area, and thus requires coordinated use of the available resources.

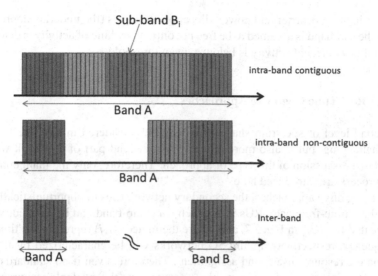

Fig. 2.5 Carrier aggregation

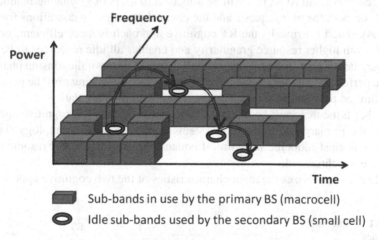

Fig. 2.6 Opportunistic sub-band allocation in NRT cognitive approaches

In NRT cognitive approaches, the main problem is making a cognitive device aware, in an autonomic manner, of the frequency sub-bands that are not being used by the primary system, thus using the transmission opportunities. The main problem here is therefore the spectrum sensing operation.

With reference to this problem, an interesting example of the NRT cognitive method is presented in [1], where the authors propose a selective method that allows the secondary user to use not only the sub-bands that were left unused by the primary system (the overlay approach), but also the sub-bands that were underutilized, by using a suitable subcarrier and power allocation strategies (the underlay approach). Indeed, the sub-band is assumed to be free not only if any kind of activity is detected, but also if any detected activity is below a given threshold.

2.4.1.2 Real Time Cognitive Approaches

The second level of spectrum sharing that can be considered in a 5G network is more challenging, but is also more efficient. The second part of this book will be devoted to a discussion of this type of approach. Therefore, only the main concepts of this process are introduced here.

In RT cognitive approaches, the secondary network tries to opportunistically use the smallest time-frequency RUs into which each sub-band can be divided, which are indicated by $RU_{i,j}$ in Fig. 2.7, and where the indices (i,j) represent the time and the frequency, respectively. Future 5G networks will be characterized by flexible and dynamic resource usage and assignment. Therefore, even if no standards exist for 5G communications, it is reasonable to assume that the available resources will be divided into small RUs[1] that will be allocated to users in a dynamic manner that depends on both the user requests and the channel propagation conditions frame by frame. As stated previously, the RT cognitive approach is more efficient, because it works with higher resource granularity and enables all idle resources to be used. However, the RT approach is also very challenging in terms of the sensing phase that must be performed on a time-scheduled basis. Therefore, sensing must be performed over a limited time period and must be repeated with high regularity.

Another issue that must be taken into account is that the RT cognitive approach requires the primary and secondary systems to use the same technology (i.e., the two systems must adopt the same time-frequency organization as the resources) and to be time-synchronized.

Table 2.1 summarizes the main characteristics of the two cognitive approaches.

Table 2.1 Cognitive approaches

	NRT	RT
Sensing period	Several frames	Scheduling period
Transmission opportunity	Sub-bands	Resource units
Technical challenges	Accurate sensing	Accurate sensing Feedback information Joint resource allocation
Spectrum efficiency	Low	High
Secondary network requirements	No	Synchronization and legacy terminal

---

[1]For example, the PRBs in LTE-A.

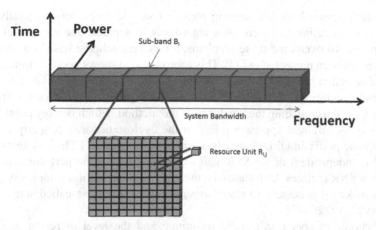

Fig. 2.7 RT approaches for spectrum opportunities

2.4.2 Spectrum Sensing

Cognitive approaches can be highly effective, but only if the knowledge of the environment is reliable, and thus require efficient spectrum sensing procedures to be used. For this reason, several spectrum sensing techniques have been proposed in the literature. In [49] in particular, an interesting overview of the different possible methods is provided, including one well-known method based on the energy detection approach. The method is blind, meaning that it does not require any knowledge of the primary system features, and is therefore particularly useful when the signal to be detected is unknown. In addition, the method can be implemented using a low complexity and low cost algorithm. To determine whether a signal is present or not in a given area, the energy detector scheme estimates the energies of the samples that are received during a dedicated period (i.e., secondary system communication is avoided during this interval) and then compares these energies with a specific threshold, which has a value that is set based on the background noise power at the receiver. If the energy is above the threshold, then another system is assumed to be present and transmitting in that area. This scheme is the easiest to implement, but it has the drawback that it only provides good performance if perfect knowledge of the noise variance is guaranteed. It thus suffers from heavy degradation in cases of noise uncertainty (i.e., low SNR regimes). This issue is widely discussed in [8, 42], where the presence of the *SNR wall* (i.e., the SNR below which we cannot reach any specified value for the false alarm and detection probability by simply increasing the observation time) is also highlighted. In addition, noise estimation is not a trivial task. The estimation process should be performed in time slots where useful signals are not present, and thus requires an initial rough sensing phase[2] to detect idle

---

[2]Using sensing strategies that work without knowledge of the noise variance.

time-slots followed by a fine sensing phase, or should be performed jointly with detection while using more complex algorithms to separate the signal and noise components. To overcome these problems, a coherent scheme based on preamble detection has been presented in [32]. This approach also shows good performance in terms of detection probability under low SNR conditions. However, like all matched filter implementations, this method requires a dedicated receiver for the specific primary user, thus reducing the flexibility of the method, which is a key point of CR technology. A different approach relies on the cyclostationarity property induced by the cyclic prefix in OFDM signals, such as those used in LTE-A systems. This method is independent of the *SNR wall* and can achieve good performance levels under low SNR regimes. Unfortunately, the high implementation complexity of this method makes it necessary to resort to sub-optimal solutions called single cycle detectors [7, 12, 20, 42].

The choice of spectrum sensing technique and the level of resources (either in terms of radio or computational resources) that are to be dedicated to sensing require careful evaluation based on the operational scenario and the CR approach used. In particular, algorithms with higher complexity and higher accuracy, such as cyclostationary detector algorithms, require sensing periods that are suitable for NRT cognitive approaches but are not suitable for RT approaches, while ED or SINR measurements are less accurate but are more flexible. In the RT case, the small cell must be able to acquire knowledge of the resource usage of the macrocell on a RU basis. The problem with that is that the physical RU is composed using only a few samples; using a single RU for sensing may therefore be insufficient, and thus would lead to inaccurate estimates (i.e., high false alarm and missed detection probabilities). However, more than one RU is usually allocated to a UE by the macrocell, and the possibility of gathering together all RUs that were allocated to a given user to perform sensing based on a wider set of samples could represent a significant improvement. The problem here is that knowledge of the scheduling information of the macrocell (i.e., which of the RUs are assigned to a given UE) cannot be available at the small cell if there is no coordination among the cells. One possible solution is represented by use of the similarity of the eigenspaces of the signals that belong to the same user [4]. The idea, which was proposed in [4], is to project the most informative eigenvector for each RU in the space represented by the main eigenvectors of the other RUs. The same user in fact has similar eigenvectors in each signal eigenspace of the RU that is allocated to them. Therefore, the correlation between the eigenvectors of the RUs for the same user is high, and it is therefore possible to identify groups of RUs for each user.

References

1. Bansal G., Hossain M.J., Bhargava V.K. and Tho Le-Ngoc (2013) Subcarrier and Power Allocation for OFDMA-Based Cognitive Radio Systems With Joint Overlay and Underlay Spectrum Access Mechanism. IEEE Transactions on Vehicular Technology 62(3): 1111–1122

2. Bartoli G., Fantacci R., Letaief K., Marabissi D., Privitera N., Pucci M. and Jun Zhang (2014) Beamforming for small cell deployment in LTE-advanced and beyond. IEEE Wireless Communications 21(2): 50–56

3. G. Bartoli, R. Fantacci, D. Marabissi, and M. Pucci (2014) Adaptive muting ratio in enhanced Inter-Cell Interference Coordination for LTE-A systems. International Wireless Communications and Mobile Computing Conference (IWCMC2014)

4. Bartoli G., Fantacci R., Marabissi D. and Pucci M. (2014) Physical Resource Block clustering method for an OFDMA cognitive femtocell system. Physical Communication

5. G. Bartoli, R. Fantacci, D. Marabissi, and M. Pucci (2014) Coordinated Scheduling and Beamforming Scheme for LTE-A HetNet Exploiting Direction of Arrival.

6. A. Bedekar and R. Agrawal (2013) Optimal muting and load balancing for eICIC. 11th International Symposium on Modeling Optimization in Mobile Ad Hoc Wireless Networks (WiOpt): 280–287

7. Bose, S. and Natarajan, B. (2012) Reliable spectrum sensing for resource allocation of cognitive radio based WiMAX femtocells. IEEE Consumer Communications and Networking Conference (CCNC):889–893

8. D. Cabric and A. Tkachenko and R. W. Brodersen (2006) Experimental study of Spectrum Sensing based on Energy Detection and Network Cooperation. Proc. of ACM TAPAS, Boston, Massachusetts, USA

9. Y. Cao, H. Xia, and C. Feng (2013) Evaluation of diverse Cell Range Expansion strategies applying CoMP in heterogeneous network. IEEE 24th International Symposium on Personal Indoor and Mobile Radio Communications (PIMRC): 1962–1966

10. M. Cierny, H. Wang, R. Wichman, Z. Ding, and C. Wijting (2013) On Number of Almost Blank Subframes in Heterogeneous Cellular Networks. IEEE Transactions on Wireless Communications 12(10): 5061–5073

11. S. Deb, P. Monogioudis, J. Miernik, and J. Seymour (2014) Algorithms for Enhanced Inter-Cell Interference Coordination (eICIC) in LTE HetNets. IEEE/ACM Transactions on Networking 22(1): 137–150

12. Derakhshani Mahsa and Nasiri-Kenari Masoumeh and Le-Ngoc Tho(2010) Cooperative Cyclostationary Spectrum Sensing in Cognitive Radios at Low SNR Regimes. Proceedings of IEEE International Conference on Communications (ICC)

13. Di Benedetto, Maria-Gabriella, Bader, Faouzi (Eds.) (2014) Cognitive Communication and Cooperative HetNet Coexistence. Series: Signals and Communication Technology 2014, XVI

14. B. Du, C. Pan, W. Zhang, and M. Chen (2014) Distributed Energy-Efficient Power Optimization for CoMP Systems With Max-Min Fairness. IEEE Communications Letters 18(6):999–1002

15. S. Fu, B. Wu, H. Wen, P.-H. Ho, and G. Feng (2014) Transmission Scheduling and Game Theoretical Power Allocation for Interference Coordination in CoMP. IEEE Transactions on Wireless Communications 13(1): 112–123

16. I. Guvenc (2011) Capacity and Fairness Analysis of Heterogeneous Networks with Range Expansion and Interference Coordination. IEEE Communications Letters 15(10): 1084–1087

17. Hamza A.S., Khalifa S.S., Hamza H.S. and Elsayed K. A (2013) Survey on Inter-Cell Interference Coordination Techniques in OFDMA-Based Cellular Networks. IEEE Communications Surveys Tutorials 15(4): 1642–1670

18. Hossain E., Rasti M., Tabassum H. and Abdelnasser A. (2014) Evolution toward 5G multi-tier cellular wireless networks: An interference management perspective. IEEE Wireless Communications 21(3):118–127

19. Hu R.Q. and Yi Qian (2014) Resource Management for Heterogeneous Networks in LTE Systems. Series: SpringerBriefs in Electrical and Computer Engineering XII

20. Huang G. and Tugnait J.K. (2011) On Cyclic Autocorrelation Based Spectrum Sensing for Cognitive Radio Systems in Gaussian Noise. 49th Annual Allerton Communication, Control, and Computing (Allerton) Conference on:501–507

21. Huang L., Guangxi Zhu and Xiaojiang Du (2013) Cognitive femtocell networks: an opportunistic spectrum access for future indoor wireless coverage. IEEE

22. U. Jang, H. Son, J. Park, and S. Lee (2011) CoMP-CSB for ICI Nulling with User Selection. IEEE Wireless Communications 10(9): 2982–2993
23. K. Kwak, H. Lee, H. W. Je, J. Hong, and S. Choi (2013) Adaptive and Distributed CoMP Scheduling in LTE-Advanced Systems. In Proc. IEEE 78-th Veh. Technol. Conf. (VTC Fall): 1–5
24. T. Lakshmana, J. Li, C. Botella, A. Papadogiannis, and T. Svensson (2013) Scheduling for backhaul load reduction in CoMP. IEEE Wireless Communications and Networking Conference (WCNC): 227–232
25. S. Lembo, P. Lunden, O. Tirkkonen, and K. Valkealahti (2013) Optimal muting ratio for Enhanced Inter-Cell Interference Coordination (eICIC) in HetNets. IEEE International Conference on Communications Workshops (ICC): 1145–1149
26. X. Li, Q. Cui, Y. Liu, and X. Tao (2012) An effective scheduling scheme for CoMP in heterogeneous scenario. IEEE 23rd International Symposium on Personal Indoor and Mobile Radio Communications (PIMRC): 870–874
27. Lien S., Chen K., Liang Y. and Lin Y. (2014) Cognitive radio resource management for future cellular networks. IEEE Wireless Communications 21(1):70–79
28. Lin Xingqin, Andrews Jeffrey G., Ratasuk Rapeepat, Mondal Bishwarup and Ghosh Amitava (2013) Carrier aggregation in heterogeneous cellular networks. IEEE International Conference on Communications (ICC): 5199–5203.
29. A. Liu, V. Lau, L. Ruan, J. Chen, and D. Xiao (2014) Hierarchical Radio Resource Optimization for Heterogeneous Networks With Enhanced Inter-Cell Interference Coordination (eICIC). IEEE Transactions on Signal Processing 62(7): 1684–1693
30. Lopez-Perez D., Guvenc I. De la Roche G., Kountouris M., Quek T.Q.S. and Jie Zhang (2011) Enhanced intercell interference coordination challenges in heterogeneous networks. IEEE Wireless Communications 18(3):22–30
31. D. Lopez-Perez, X. Chu, and I. Guvenc (2012) On the Expanded Region of Picocells in Heterogeneous Networks. IEEE Journal of Selected Topics in Signal Processing 6(3): 281–294, June 2012
32. Mahmoud A Abdelmonem and Mohammed Nafie and Mahmoud H Ismail and Magdy S El-Soudani (2012) Optimized spectrum sensing algorithms for cognitive LTE femtocells. EURASIP Journal on Wireless Communications and Networking
33. P. Marsch and G. Fettweis (2011) Static Clustering for Cooperative Multi-Point (CoMP) in Mobile Communications. IEEE International Conference on Communications (ICC): 1–6
34. G. Morozov and A. Davydov (2013)CS/CB CoMP scheme with semi-static data traffic offloading in HetNets. in Proc. IEEE 24-th Int. Symp. Pers. Indoor Mobile Radio Commun. (PIMRC): 1347–1351.
35. Naranjo Juan Diego, Bauch Gerhard, Saleh Abdallah Bou, Viering Ingo and Halfmann, Ruediger (2013) A Dynamic Spectrum Access Scheme for an LTE-Advanced HetNet with Carrier Aggregation. Proceedings of 2013 9th International ITG Conference on Systems, Communication and Coding (SCC): 1–6
36. Ngo, Duy Trong, Le-Ngoc, Tho (2014) Architectures of Small-Cell Networks and Interference Management. Series: SpringerBriefs in Computer Science XII.
37. A. Papadogiannis, D. Gesbert, and E. Hardouin (2008) A Dynamic Clustering Approach in Wireless Networks with Multi-Cell Cooperative Processing. IEEE International Conference on Communications (ICC): 4033–4037
38. Samsung (2010) System Performance of Heterogeneous Networks with Range Expansion. 3GPP R1-101203, Tech. Rep.
39. Sangmi Moon, Bora Kim, Saransh Malik, Cheolwoo You, Huaping Liu, Jeong-Ho Kim and Intae Hwang (2015) Interference Management with Cell Selection Using Cell Range Expansion and ABS in the Heterogeneous Network Based on LTE-Advanced Wireless Personal Communications 81(1):151–160
40. Saquib N., Hossain E., Long Bao Le and Dong In Kim (2012) Interference management in OFDMA femtocell networks: issues and approaches. IEEE Wireless Communications 19(3): 86–95

41. B. Soret, H. Wang, K. Pedersen, and C. Rosa (2103) Multicell cooperation for LTE-advanced heterogeneous network scenarios. IEEE Wireless Communications 20(1): 27–34
42. Tani A. and Fantacci R. (2010) A Low-Complexity Cyclostationary-Based Spectrum Sensing for UWB and WiMAX Coexistence With Noise Uncertainty. IEEE Transactions on Vehicular Technology 59(6): 2940–2950
43. P. Tian, H. Tian, J. Zhu, L. Chen, and X. She (2011) An adaptive bias configuration strategy for Range Extension in LTE-Advanced Heterogeneous Networks. IET International Conference on Communication Technology and Application (ICCTA 2011): 336–340.
44. S. Vasudevan, R. Pupala, and K. Sivanesan (2013) Dynamic eICIC: A Proactive Strategy for Improving Spectral Efficiencies of Heterogeneous LTE Cellular Networks by Leveraging User Mobility and Traffic Dynamics. IEEE Transactions on Wireless Communications 12(10): 4956–4969
45. Y. Wang and K. Pedersen (2011) Time and Power Domain Interference Management for LTE Networks with Macro-Cells and HeNBs. IEEE Vehicular Technology Conference (VTC Fall): pp. 1–6
46. Xiao Y., Yuen C., Di Francesco P. and DaSilva L. A. (2013) Dynamic spectrum scheduling for carrier aggregation: A game theoretic approach. IEEE International Conference on Communications (ICC): 2672–2676.
47. Z. Xu, C. Yang, G. Li, Y. Liu, and S. Xu (2014) Energy-Efficient CoMP Precoding in Heterogeneous Networks. IEEE Transactions on Signal Processing 62(4): 1005–1017
48. Ye Qiaoyang, Beiyu Rong, Yudong Chen, Al-Shalash M., Caramanis C. and Andrews, J.G. (2013) User Association for Load Balancing in Heterogeneous Cellular Networks. IEEE Transactions on Wireless Communications 12(6):2706–2716
49. Yucek, T. and Arslan, H. (2009) A survey of spectrum sensing algorithms for cognitive radio applications. IEEE Communications Surveys Tutorials 11(1):116–130
50. J. Zhao, T. Quek, and Z. Lei (2013) Clustering method for CoMP with limited backhaul data transfer using convex relaxation. IEEE International Conference on Communications (ICC): 5398–5403

Part II
Congitive Beamforming Approaches for HetNets

Chapter 3
Cognitive HetNets

3.1 Introduction

The deployment of two overlapped layers of cells (i.e., macrocells and small cells) leads to a requirement for self-organization capabilities to enable efficient management of the inter-cell interference.

In 4G networks, the introduction of small cells focuses mainly on coordinated approaches, where the small cell is deployed directly by the operator, and it is therefore possible to adopt coordinated resource allocation strategies to prevent inter-cell interference. However, in future 5G networks, the number of small cell nodes is expected to increase significantly, and many user-deployed small cells will be used in different environments, including homes, small offices and enterprises. In fact, it has been estimated that most data traffic in the future will originate from indoor environments, where small cells will be able to reach the expected data rates and satisfy increasing user bandwidth requests. In addition, ad hoc deployment of small cells will enable more cost-effective cell densification to be achieved, thus reducing the costs for RF planning, site acquisition and efficient backhauls. However, in this scenario, coordination among the macrocells and the small cells for resource assignment will be either infeasible or impossible (because of the network delay and the signaling overheads), and this has led to the concept of cognitive HetNets [8].

The primary system, i.e., the macrocell, has a higher priority than a small cell for resource usage. In addition, small cells should handle data transmission without affecting primary system reception. To achieve this goal, each small cell must therefore be able to sense the environment to acquire knowledge of the primary system transmissions and then adapt its own transmission/reception processes using opportunistic and dynamic resource allocation schemes combined with advanced signal processing methods.

© The Author(s) 2015
D. Marabissi, R. Fantacci, *Cognitive Interference Management in Heterogeneous Networks*, SpringerBriefs in Electrical and Computer Engineering,
DOI 10.1007/978-3-319-20191-7_3

As stated in Chap. 2 (Sect. 2.4.1), cognitive approaches can be divided into two main groups, depending on the time granularity of their operations. In particular, NRT approaches operate on a long time scale that allows dynamic frequency partitioning that depends on the averaged traffic load. Conversely, RT approaches operate by adapting the allocated resources to instantaneous traffic requests on a scheduled time basis. In addition, RT approaches operate on a single RU, while NRT approaches operate over the whole frequency band.

This chapter focuses on the RT approaches, which represent very promising solutions because of the specific features that they offer; however, at the same time RT approaches present many challenging issues to be solved such that their use in 5G systems becomes viable.

The main goals of RT cognitive approaches are:

1. Awareness in real time of the use of spectrum resources by the macrocell.
2. Making quick and efficient access decisions for the secondary devices.

The first goal depends on the need to adapt the transmission/reception each scheduled time, while the second goal depends on the need to select and use an efficient (e.g., optimal) resource allocation decision to minimize any interference towards the primary system without affecting the performance of the secondary devices.

The resource allocation schemes benefit from the tremendous flexibility given by multiuser diversity; when working on a RU basis, different secondary users have different spectrum opportunities available. Indeed, depending on their positions, the secondary users receive and produce different levels of interference from/towards the primary users. This means that if the k-th macrocell user equipment, MUE_k, is communicating using the (i,j)-th resource $RU_{i,j}$,[1] then the x-th nearby small cell user equipment, SUE_x, senses the $RU_{i,j}$ as being occupied, but another SUE, SUE_z, which is located far away from MUE_k, senses $RU_{i,j}$ as being free and can thus use it for communication. To use this multiuser diversity property, the spectrum sensing must perform joint spatial-temporal resource detection.

Awareness of the macrocell resource usage can be achieved in two different ways: by receiving the scheduling information from the MBS or by actively sensing the environment.

In the first case, a limited level of signaling exchange among the BSs is required, while in the second case, the information is acquired using sensing procedures without any type of information exchange. The first method is not completely cognitive because there is only limited cooperation between the BSs, but the information exchange represents only one-sided cooperation, and interference management mainly relies on cognitive approaches. In addition, use of the scheduling information sent by the macrocell has the drawback that the multiuser diversity remains unused.

---

[1] (i,j) denotes the i-th time slot and the j-th frequency (see Sect. 2.4.1.2).

Indeed, when a resource is used by the primary system is always considered busy, it cannot be used by the secondary system or by a SUE located far away from the primary system, and thus can operate without interference.

The second method is completely cognitive but presents more challenges in the sensing phase, not only in terms of the accuracy of the results, but also in terms of the latency between acquisition of the context awareness and its use and the channel reciprocity between the UL and the DL. In some cases, hybrid techniques can represent a viable solution whenever limited knowledge of the network is available, and therefore only partial coordination among the cells is possible.

Another issue to be considered is that the sensing differs between the UL and the DL. The BS of a cognitive secondary system listens to the environment to acquire knowledge about the UL transmissions of the primary system. This information can be used to allocate resources in a successive UL transmission phase. In contrast, the DL knowledge must be acquired by the terminals during a given time interval and sent back to the access point that performs resource allocation in the subsequent DL frame. This procedure introduces a certain latency between acquisition of the information and its use, which must be accurately evaluated in accordance with the working speed of the scheduler. In particular, this could be critical for TDD systems, where the cognitive terminal must wait for the appropriate UL subframe before sending back the information to the access point. Despite this latency, TDD-based systems are attractive for use of RT cognitive methods because channel reciprocity is applicable for these systems, and allows channel estimates that were performed in the DL to also remain valid for the UL, and vice versa.

3.2 Cognitive Solutions for 5G Systems

This section deals with some possible methodologies for application to cognitive HetNet communications.

3.2.1 Interference Scenarios

In HetNets, the small cells are low power nodes, and the interference level that a small cell can generate is therefore lower than the interference generated by a macrocell. However, when a MUE is very close to a small cell or is within the coverage area of a small cell but is not permitted to connect to it (i.e., a closed access small cell), the level of interference on the MUE can be very high to the point of being disruptive. In this case, it is possible to use cognitive approaches to prevent small cell transmissions or limit the power emitted by the resources assigned to the MUEs in the small cell coverage area.

Conversely, in the case of the UL communications, the MUEs can cause severe interference at the SBS when they are located far away from the MBS because of

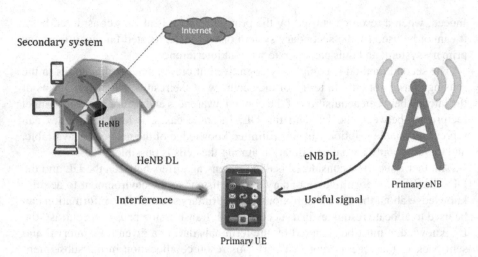

Fig. 3.1 DL communication interference scenario

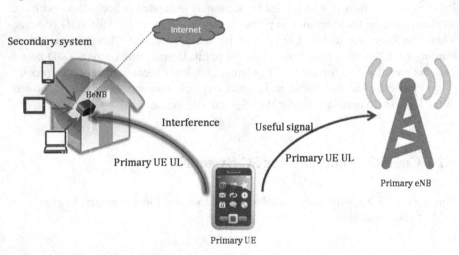

Fig. 3.2 UL communication interference scenario

their high transmission power. In this case, the signal received by the SBS from the MUE can be significantly higher than the signal received by a SUE transmitting using the same resources.

The two scenarios are illustrated in Figs. 3.1 and 3.2.

Although the UL and DL scenarios are different, similar policies can be applied to both links. In general, the goal is to maximize the secondary system performance (e.g., in terms of throughput) with certain constraints placed on the interference level generated towards the macrocell system. Some examples of this are reported in the next section.

3.2.2 Interference Avoidance

Most papers on this subject in the literature are concerned with cognitive small cells that look for resources that are not being used by the macrocell. This means not only blank macrocell resources, but also resources that have been allocated to distant MUEs, which are interference-free as a result. For example, the method proposed in [11] first detects channel occupation by estimation of the energy in the UL sub-channels, and then allocates the sub-channels with the lowest interference signatures to the small cell users. The hypothesis here is that the same resource scheduling process is used in both UL and DL transmissions. However, it is more likely that the DL and the UL are characterized by asymmetric traffic, and thus adopt different resource allocation policies. Therefore, UL sensing cannot be used for DL allocation. One alternative approach [8] is based on a hybrid sensing scheme, in which the scheduled MBS information is available at the SBS and can be used to increase the spectrum sensing accuracy. In this way, the secondary system finds more spectrum opportunities by identifying nearby macrocell users. The resources are then assigned using an iterative procedure based on the Gale-Shapley theorem that was originally proposed in [2] to solve the stable marriage problem (a well-known problem in mathematics and computer science), which is defined as follows [2]:

> A certain community consists of n men and n women. Each person ranks those of the opposite sex in accordance with his or her preferences for a marriage partner. We seek a satisfactory way of marrying all the members of the community. [..] we call a set of marriages unstable [..] if under it there are a man and a woman who are not married to each other but prefer each other to their actual mates.

In a more general sense, the problem to be solved consists of finding a stable match, i.e., a mapping of the elements of one set to the elements of the other set, given certain preferences for each element. In this case, the two sets are represented by the users and the resources.

Inter-layer interference can also be limited or prevented using optimal power allocation, and using underlay cognitive approaches in particular. In [14], the SBS senses both the UL and DL of the macrocell to be aware of both the resource occupancy and the nearby MUEs. In particular, UL sensing is used to detect the presence of the MUEs and to estimate the path loss between an SBS and an MUE. Similarly, DL sensing is used to determine whether a resource is being used by the macrocell (if the RSS exceeds a suitable threshold value) or not. The algorithm then adapts the power on each resource element to maximize the achievable small cell throughput and fulfill the MUE outage constraints. In contrast, in [25], a small cell is allowed to transmit its data using either idle or busy resources by paying suitable *prices* to the macrocell. To describe this process in greater detail, the authors proposed a method where the macrocell protects itself by setting a maximum aggregate interference constraint, and then makes a profit by pricing the interference that arises from SUEs. Two different prices are set, depending on whether the resource is idle or busy. Given these prices and the sensing results, the small cells then choose their transmission power using game theory, which is widely adopted to solve resource allocation problems in cognitive networks. A brief discussion on game theory is provided in Sect. 3.2.2.1. Two other examples of game theory applied in the context of cognitive HetNets are presented in [20] and [7]. In particular, in [20], a supermodular game is used to solve the joint subcarrier and power allocation problem. Each SUE randomly selects a certain number of subcarriers, depending on its needs. These new subcarriers are compared with those selected at previous iterations in terms of the payoff[2] (i.e., a utility function), which in this case is represented by the RSS difference between one specific player (i.e., a SUE) and the other players on each subcarrier. If there are newly selected subcarriers with utilities that are higher than those of the previous subcarriers, the SUE accepts these newly selected subcarriers by removing the others. This process continues until each SUE reaches equilibrium (see Sect. 3.2.2.1). In contrast, in [7], the SBS performs the spectrum sensing to detect the free resources that are not being used by the macrocell, and the resources are then allocated using a *regret-matching* procedure from game theory. Each SUE learns about the *regret* from its previous actions to acquire a specific RU, and aims to minimize its *regret* value at the next time instant. The power is then allocated using an optimal water-filling approach. Finally, an interference avoidance scheme is used to limit the interference in the case of sensing errors.

The combined use of the subcarrier and power allocation techniques is also proposed in [1], where, after the initial power distribution, subcarrier allocation is performed using the Hungarian method, which is a combinatorial optimization algorithm that solves the problem in polynomial time [6]. Specifically, by assuming that a numerical score can be associated with the performance of each user on each resource, the Hungarian method allows optimal assignment of n resources to n users. The goal is to maximize the numerical scores, which were derived in [1] by taking the throughput of the secondary users and a constraint on the SINRs of the primary UEs into account.

In another approach, the subcarrier-power allocation problem is solved in [5] via a distributed two-step algorithm. After an initialization phase, during which the spectrum sensing, cell association and CSI estimation steps are all performed, optimal subcarrier allocation is carried out with the goal of maximizing the throughput for each small cell by using the spectrum sensing information that was acquired by the SBS. The power allocation problem is then solved using an iterative geometric programming approach, which is a general mathematical optimization tool. Similarly, the algorithm proposed in [17] first performs subcarrier allocation based on the channel gains and the interference levels towards the MUEs, and the power is then optimally distributed among the subcarriers with the aim of satisfying the rate requirements of the heterogeneous users. In this case, several practical limitations, including imperfect spectrum sensing, limited transmission power, and the various traffic demands of the secondary users are all taken into account.

---

[2]The payoff in a game represents the motivations of the players. It may represent profit or utility, or may simply rank the desirability of the possible outcomes.

3.2.2.1 Game Theory

Game theory is a discipline that has widespread popularity in applied mathematics, in which the main scope is the description and analysis of interactive decision-making strategies. Game theory specifically consists of a set of analytical tools that are used to predict the outcome of conflict and cooperation interactions among intelligent rational decision elements [9].

A game involves three basic sets of elements:

* The players;
* The actions;
* The payoff;

where the players act as the decision makers.

Usually in game theory we can distinguish between non-cooperative and cooperative games. In a non-cooperative game, the players all act independently. Conversely, a cooperative game is a game where subsets of the players (which are usually named coalitions) may enforce cooperative behavior, and thus the game is a competition between these coalitions of players rather than between the individual players. An important difference between the two types of game described above is the availability of the payoffs for the other players. In cooperative games, users in a subset are aware of what the other users do. Conversely, in non-cooperative games, this capability is not allowed. In fact, communication among the players is usually enabled in cooperative games, but is not enabled in non-cooperative games. A cooperative game is set up by specifying a value for each coalition. This function describes how much a set of players can gain by forming a specific coalition. In the case of non-cooperative games, each player can be considered to select the most suitable strategy to achieve the best possible utility for their purposes. Of the two types of game, non-cooperative games aim to model situations to the finest detail and produce accurate results, while cooperative games are more interested in the game in general.

Examples of games that can be retrieved from daily life are all conventional games of strategy, including chess and poker, strategic negotiations performed in the purchase of items or objects, and daily group decision-making processes, such as decisions on where to go to dinner with friends or what show to see.

However, a clear distinction should be drawn between a game, which must involve multiple decision makers, and an optimization problem, which involves only a single decision maker. A game model is usually appropriate only in scenarios where we can reasonably expect that the decisions of each player will affect outcomes that are relevant to the other players. For example, a single consumer buying goods at a store is usually performing an optimization process by pursuing maximum satisfaction with the items purchased by taking the available budget into account. Conversely, a game model is applicable when we consider a store being involved in a pricing strategy with other stores in the same district.

Modern game theory began with the contributions of von Neumann [15, 16] that originated cooperative game theory, which analyzes optimal strategies for groups of individuals by presuming that they can establish unbreakable agreements between them about the strategies to be followed. In this context, the contribution of Nash constituted a breakthrough. In [10], he developed a criterion, known as the Nash equilibrium, that is applicable to a wider variety of games than the criterion that was proposed in [16]. Thanks to Nash's contribution, it became possible to begin analysis of non-cooperative games.

Recently, game theory received considerable attention for its use in the solution of decentralized resource allocation problems in wireless communication networks, and in particular in cognitive HetNets, where there is no central unit for resource allocation. In this scenario, each SUE must dynamically select the resources that can be shared with the MUE, limiting the interference level for the MUE to less than a desired level while providing the desired throughput for the SUE.

In the HetNets case, as outlined in the literature [7, 20, 25], the players are most often the small cell users. The actions are generally the strategies related to reduction of the negative effects of inter-cell interference, while the payoff represents the utility for each SUE. In general, in wireless systems, possible actions may include the choice of a modulation scheme, coding rate, protocol, flow control parameter, transmission power level, or any other factor that is under the control of a specific user. For example, in [20], the payoff is the difference between the RSS of a given player and that of the other players.

3.3 Cognitive Small Cells with Multiple Antennas

Use of multi-antenna technologies represents an additional opportunity to allow co-channel frequency allocation of primary and secondary systems using information related to the positions of the MUEs and SUEs.

One particularly useful approach uses beamforming at the small cell transmitter to provide spatial separation of the interfering signals. Each small cell is equipped with an antenna system composed of correlated elements, from which the radiated beam can be modeled. The steering accuracy and null beam selectivity grow linearly with the number of elements used in the antenna system.

The prospect therefore exists of the 3D beamforming concept being used in 5G wireless networks to allow the beam to be steered in both the azimuth and elevation direction for full spatial reuse.

The use of beamforming at the secondary transmitter assumes that the transmitter performance is maximized, while the interference directed towards the primary receiver is minimized [21, 23]. This operation, which is also known as CB, requires the solution of optimization problems to determine optimal precoding vectors and power allocation strategies based on complex numerical solutions. CB is based on knowledge of all the propagation channels over which the secondary transmitter can interfere with the primary receiver. This is impractical in real scenarios, because the two systems operate independently, and the primary system ignores the presence of the secondary system. It is therefore necessary to resort to sub-optimal solutions that can work based on partial knowledge of the CSI or based on certain information exchange between the primary and secondary networks. As an example, a solution to this problem in a TDD system has been proposed in [24], where, by using the channel reciprocity, the secondary BS estimates an effective interference channel through periodic observation of the primary system transmissions. The proposed learning-based CB scheme avoids the overhead required for the primary UE to estimate the channel from the secondary BS, and then feeds that back to the secondary BS. This makes the CB approach more practically applicable in real systems.

In [12], a CB scheme that uses non-perfect channel knowledge is proposed for the specific cognitive HetNet environment. The SUEs estimate the channel parameters using suitable reference signals and report these parameters to the SBS. Using this information, the SBS can adjust the transmission power to provide the desired SINR for the cell edge SUEs and also model the transmission beams to maximize the total sum-rate capacity of the system.

Multi-antenna systems also introduce another resource dimension: the DoA. Specifically, a small cell beam can be steered in a direction that maximizes the information signal, while nulls can be placed in the DoA of each of the primary system signals.

In [18], estimation of the DoA of the primary system is introduced as a new spectrum opportunity. In particular, a method based on the ratio of the maximum eigenvalue to the minimum eigenvalue of the covariance matrix of the received signal is used to detect the presence of the primary signal. Then, the MuSiC algorithm (see Sect. 4.3.1) is applied to primary signal DoA estimation. In [18], the focus is placed entirely on DoA estimation, and the way in which this information can be used in a cognitive system is not taken into account. Conversely, in the scheme proposed in [19], the secondary users insert a null in the direction of the primary BS, and thus avoid interference on the primary UL, but the DoA estimation problem is not considered in this case, and this information is supposed to be known perfectly.

However, in general, knowledge of the DoA of multiple signals received from multiple UEs (both SUEs and MUEs) can represent a challenging problem in a multipath propagation channel and is strongly dependent on the number of antenna elements used at the receiver. The next chapter will discuss a method for cognitive HetNets based on DoA estimation and ZFBF that takes the problems that arise under the assumption of actual propagation conditions into account.

CB methods can also be combined with opportunistic resource allocation algorithms that assign each RU to a SUE that is sufficiently distant from the corresponding MUE, thus allowing the interference to be minimized using suitable beamforming weights.

Figure 3.3 shows how the small cell can assign the resources used by MUE-2 to SUE-1, but cannot assign the resources used by MUE-1 because the beamforming process is unable to separate these users. In this method, knowledge of the DoA can be acquired by the SBS in the UL and then used either for UL reception or for DL transmission.

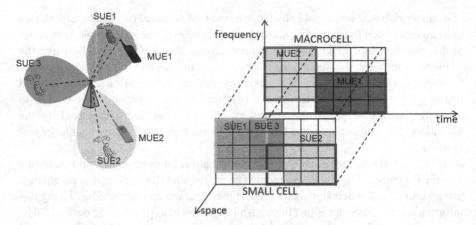

Fig. 3.3 Spatial separation of interfering UEs

Most examples of the joint use of beamforming and resource allocation that are available from the literature are based on coordination between the primary and secondary systems. In contrast, cognitive approaches are proposed in [4] and [22] that consider the joint use of suitable resource allocation schemes and ZFBF. In these approaches, beamforming weights are selected to reduce the mutual interference among the different systems by taking advantage of the spatial separation between the users. These methods are both based on semi-orthogonal user selection. The methods require knowledge of the CSI of the interference links. However, this type of approach can be too complex for use in cognitive scenarios, where the primary system ignores the presence of the secondary system. These methods also require inversion of the channel matrix of the users to calculate the beamforming weights. The next chapter presents a joint beamforming and resource allocation scheme for an LTE-A system that uses the angular information of the primary and secondary UEs rather than the CSI of the interference links.

When multiple antenna systems are available, another interesting approach that can be used to mitigate interference is based on the interference alignment principle. The goal of this approach is to select precoding matrices that allow alignment of the aggregated interference from multiple sources into a lower dimensional subspace, which means that it can easily be filtered out by the non-intended receiver by sacrificing certain signal dimensions. Examples of the application of such an approach are presented in [3, 13].

In theory, when using IA, it is possible to remove the interference completely, even if the filtering process at the non-intended receiver removes the signal energy in the interference subspace and reduces the multiplexing gain. IA is also an interesting theoretical approach, but has several practical issues that must be solved for the method to be used effectively in future cellular systems. Among these issues, two of the major challenges include the need for the transmitter to know the CSI of the non-intended receiver and the computational complexity of the precoding vectors.

References

1. Forouzan, N. and Ghorashi, S.A (2014) New algorithm for joint subchannel and power allocation in multi-cell OFDMA-based cognitive radio networks. IET Communications 8(4):508–515

2. Gale D. and Shapley L.S. (1962) College Admissions and the Stability of Marriage. American Mathematical Monthly 69: 9–14

3. Guler B. and Yener, A. (2014) Selective Interference Alignment for MIMO Cognitive Femtocell Networks. IEEE Journal on Selected Areas in Communications 32(3): 439–450

4. Hamdi K., Zhang Wei and Letaief K. (2009) Opportunistic spectrum sharing in cognitive MIMO wireless networks. IEEE Transaction on Wireless Communications 8(8):4098–4109

5. Khan, F.H. and Young-June Choi (2012) Joint subcarrier and power allocations in OFDMA-based cognitive femtocell networks. 18thAsia-Pacific Conference on Communications(APCC): 812–817

6. Harold W. Kuhn (1955) The Hungarian Method for the assignment problem. Naval Research Logistics Quarterly 2:83–97

7. Lai Wei-Sheng, Chiang Muh-En, Lee Shen-Chung and Lee Ta-Sung (2013) Game theoretic distributed dynamic resource allocation with interference avoidance in cognitive femtocell networks. IEEE Wireless Communications Network Conference (WCNC):3364–3369

8. Li Huang, Guangxi Zhu and Xiaojiang Du (2013) Cognitive femtocell networks: an opportunistic spectrum access for future indoor wireless coverage. IEEE Wireless Communications 20(2):44–51

9. Mackenzie Allen B. and DaSilva Luiz A. (2006) Game Theory for Wireless Engineers. Synthesis Lectures on Communications 1(1):1–86

10. Nash J. F.Jr. (1950) Equilibrium points in n-person games. Proceedings of the National Academy of Sciences of the United States of America 36(1): 48–49

11. Oh D.C., Lee H.C. , and Lee Y.H. (2010) Cognitive radio based femtocell resource allocation. Proc. Int. Conf. Inf. Commun. Technol. Convergence (ICTC):274–279

12. Oh D.C., Lee H.C. , and Lee Y.H. (2011) Power control and beamforming for femtocells in the presence of channel uncertainty. IEEE Transaction on Vehicular Technology 60(6): 2545–2554

13. Rihan M., Elsabrouty M., Muta O. and Fumkawa H. (2014) Iterative interference alignment in macrocell-femtocell networks: A cognitive radio approach. 11th International Symposium on Wireless Communications Systems (ISWCS):654–658

14. Sun Daolong, Zhu Xinning, Zeng Zhimin and Wan Shaohua (2011) Downlink power control in cognitive femtocell networks. International Conference on Wireless Communications and Signal Processing (WCSP):1–5

15. v. Neumann J. (1928) Zur Theorie der Gesellschaftsspiele. Mathematische Annalen 100(1):295–320

16. v. Neumann J. and Morgenstern O. (1944) Theory of Games and Economic Behavior (60th Anniversary Commemorative Edition) 60th Anniversary Commemorative edition Princeton university

17. Wang Shaowei, Zhou, Zhi-Hua, Ge Mengyao, Wang Chonggang (2013) Resource Allocation for Heterogeneous Cognitive Radio Networks with Imperfect Spectrum Sensing. IEEE Journal on Selected Areas in Communications 31(3):464–475

18. J., Fu Z. and Xian H. (2010) Spectrum sensing based on estimation of direction of arrival. International Conference on Computational Problem-Solving (ICCP): 39–42

19. Yaacoub E. and Dawy Z. (2011) Enhancing the performance of OFDMA underlay cognitive radio networks via secondary pattern nulling and primary beam steering. IEEE Wireless Communications and Networking Conference (WCNC): 1476–1481.

20. Yilmaz Mustafa Harun, Abdallah Mohamed M., Qaraqe Khalid A. and Arslan Huseyin (2014) Random subcarrier allocation with supermodular game in cognitive heterogeneous networks. IEEE Wireless Communications and Networking Conference (WCNC):1450–1455

21. Yiu S., Chae C.-B., Yang K. and Calin D. (2012) Uncoordinated Beamforming for Cognitive Networks. IEEE Transaction on Communications 60(5): 1390–1397
22. Yoo T. and Goldsmith, A. (2006) On the optimality of multiantenna broadcast scheduling using zero-forcing beamforming. IEEE Journal on Selected Areas in Communications 24(3):528–541
23. Zhang R. and Liang Y.C.(2008) Exploiting multi-antennas for opportunistic spectrum sharing in cognitive radio networks. IEEE Journal on Selected Topics Signal Process 2(1): 88–102
24. Zhang R.; Gao F. and Liang Y.C. (2010) Cognitive beamforming made practical: Effective interference channel and learning-throughput tradeoff. IEEE Transactions on Communications58(2):706–718
25. Zhou Qiaoqiao, Chen Zhong, Gao Feifei, Li J.C.F. and Lei Ming (2014) Pricing and power allocation in sensing-based cognitive femtocell networks.IEEE International Conference on Communications (ICC)

Chapter 4
Cognitive Resource Allocation with Beamforming

4.1 Introduction

This chapter focuses on the future prospects for HetNets through a discussion of inter-layer interference management based on combined use of beamforming and DoA estimation. In particular, the *underlay cognitive* network paradigm is discussed here, where a small cell can transmit simultaneously over the spectrum that has been allocated to the macrocell without affecting its reception. Consequently, the small cell must have cognitive capabilities to learn from its environment and adapt its transmissions to prevent interference with the macrocell DL communications by using the spatial dimension. This chapter discusses the case of a small cell equipped with a multiple antenna system and its ability to estimate the DoAs of signal replicas received from the different macrocell UEs by using suitable sensing intervals in the uplink frame. In addition, the advantages that arise by assuming that the small cell transmissions are performed using a ZFBF algorithm to reduce the interference directed towards the MUEs are investigated. The influence of the propagation conditions relative to specific operational scenarios on the performance of each solution considered is investigated. In particular, this chapter analyzes the system behavior when multipath components with a very large angular spread are generated by the propagation channel while multiple MUEs access the channel simultaneously, which results in the number of received signals being higher than the number of antennas. This challenges the capabilities of either the DoA estimation or beamforming algorithms. Consequently, the approach presented here operates on a RU basis. This allows limitation of the number of received signals to the number of multipath components without considering the number of UEs. Also, the second part of the chapter highlights the fact that operation on the RU basis brings additional improvements in terms of throughput, which can be achieved by combined use of the interference reduction scheme considered here with suitable resource allocation policies.

© The Author(s) 2015
D. Marabissi, R. Fantacci, *Cognitive Interference Management in Heterogeneous Networks*, SpringerBriefs in Electrical and Computer Engineering, DOI 10.1007/978-3-319-20191-7_4

The chapter first outlines the system model of interest in Sect. 4.2. Then, Sect. 4.3 describes the interference suppression scheme that is integrated with suitable user selection and resource allocation policies in Sect. 4.4. Some numerical results are provided to demonstrate the performance of the method and its efficiency.

4.2 System Model

An indoor cognitive small cell (also called HeNB) is overlaid on a macrocell that shares the same frequency resources, as shown in Fig. 4.1. Both cells are modeled according to the 3GPP LTE-A standard [17]. Also, to take the worst possible interference conditions into account, the two systems operate in FDD mode, with the same carrier frequency, $f_0 = 2\,\text{GHz}$, and the same bandwidth.

Following a CR approach, the macrocell and the small cell represent the primary system and the secondary system, respectively. Therefore, the primary UEs that are located within the small cell coverage area receive the signal from the eNB with either partial or complete overlapping of the HeNB signal and the consequent performance degradation. Attention here is focused on the DL communications, but the analysis is presented in general terms, and can be extended to UL communications to reduce the interference of the MUEs directed towards the HeNB.

Fig. 4.1 Scenario considered in this chapter

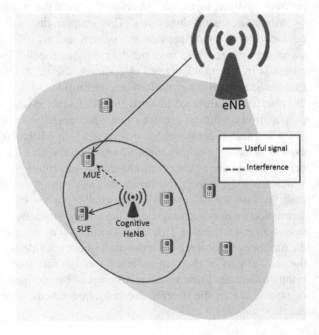

The system is based on the OFDMA transmission scheme, where the resources are organized using a time-frequency grid, and are uniquely assigned to the users. The smallest RU that can be allocated to any user is a PRB, which consists of 12 subcarriers over 7 consecutive OFDM symbols. The number of PRBs per slot, J, depends on the system bandwidth (i.e., $6 \leq J \leq 100$), which varies in the [1.25– 20.0] MHz range.

In the scheme described here, the small cell must be able to perform both DoA estimation and digital steering. Therefore, the system is equipped with a linear array of antennas, consisting of L equally spaced elements. The number of antennas has been selected to achieve a reasonable trade-off between the device dimensions and the system performance. It is well documented in the literature that as the number of antenna elements increases, the system performance also improves. In particular, DoA estimation becomes increasingly accurate and the number of resolvable directions grows linearly with the number of antennas and the beam directivity (or the null selectivity). However, a high number of antenna elements also leads to higher computational loads and higher antenna dimensions.

If we consider the array elements to be uniformly spaced with spacing $d = \lambda/2$ and $f_0 = 2$ GHz, then the total space, D, required to deploy L antennas is

$$D = \frac{(L-1)c}{2f_0} \tag{4.1}$$

where c is the speed of light. This means that each antenna element requires a space of 7.5 cm. In the system of interest here, the small cell is provided with $L = 4$ antennas, which leads to a total dimension $D = 22.5$ cm. This choice allows accurate beamforming operations to be performed, as detailed in [6], with device dimensions that are similar to those of other household devices, e.g., Wi-Fi access points. A higher number of antennas would lead to improved performance, and this aspect may be of special interest for systems that operate in higher frequency bands.

In addition, in the discussion of possible solutions for the interference suppression problem that forms the subject of the next section, multipath propagation effects are taken into account by using the tapped-delay-line model that was defined in the ITU-R M.1225 Recommendation [12]. Two different PDPs are considered here:

- Outdoor-A for the eNB – MUEs links;
- Indoor-A for the HeNB – MUEs and HeNB – SUEs links.

4.3 Adaptive Angular Interference Suppression

This section illustrates a method used to reduce HeNB interference on MUEs during DL reception [1, 3]. This method is structured in two phases. The small cell performs a first sensing phase, during which it identifies the DoAs of the primary user signals. In the second phase, the HeNB calculates suitable beamforming weights to modify the radiation pattern during transmission to place zeros such that they correspond with the estimated MUE directions. However, for actual implementation of this scheme, it is necessary to account for the fact that any DoA estimation algorithm can detect up to $L - 1$ DoAs, and that the ZFBF algorithm can place $L - 1$ nulls in the desired directions. This problem was first studied in [1], where the authors considered an OFDM system (where all subcarriers are allocated to a single user in a given time period) with a variable number of received signals M. The results, which are reported in Sect. 4.3.3, show that the interference suppression method also performs well when the number of multipath replicas is slightly higher than $L - 1$. However, this system does not take the actual LTE-A access scheme (i.e., OFDMA), where multiple users are simultaneously present, into account. In addition, more than one UE is usually present in the small cell area, and each UE is characterized by a PDP composed of multiple replicas. Therefore, the number of received signals may be significantly higher than $L - 1$ (the number is given by the number of replicas multiplied by the number of UEs). To solve this problem, a system working on a PRB-wise basis is proposed in [3]. Based on the assumption that each PRB is only allocated to a single user, performing DoA estimation and ZFBF on each PRB enables estimation of $L - 1$ DoAs for each PRB and thus allows the UE signals to be separated in reception and transmission. The number of received signals thus only depends on the PDP, as in the case studied in [1]. The problem with this is that the snapshot[1] used to estimate the DoA is composed only of the subcarriers of one PRB (i.e., 12 subcarriers in LTE-A) and thus reduces the estimation accuracy. This impairment can be compensated for by grouping of the PRBs that belong to the same UE, thus increasing the snapshot. Two possible methods that can be used to perform the grouping process are based on use of the knowledge of the macrocell scheduling information or the similarity of the eigenvectors [2]. The performance improvement that can be achieved by PRB grouping is discussed in Sect. 4.3.4.

4.3.1 Sensing Phase: DoA Estimation

During the sensing phase, the HeNB estimates the DoAs of the UEs in its coverage area.[2] For this purpose, the HeNB schedules specific OFDM symbols in each subframe in the UL transmissions. As demonstrated in [10], the UL and DL spatial information is strongly correlated, and it is therefore reasonable to use UL estimates for the DL transmissions. DoA estimation is performed for each PRB to obtain the DoAs of all F SUEs and N MUEs that are present in the small cell coverage area. Assuming that the m-th MUE is transmitting on the j-th PRB, and that the propagation channel is characterized by M multipath components, then the n-th

---

[1]The number of samples used by the estimation algorithm.

[2]An initial phase of spectrum sensing is performed to detect the presence of the MUEs.

sample of the signal that is received by the l-th antenna element of the HeNB can be expressed as

$$r^m_{l,j}[n] = \sum_{q=1}^{M} x^m_j[n - \tau^m_j(q)]h^m_j(q)e^{j[\pi(l-1)\sin(\theta^m_j(q))]} + v_l[n] \qquad (4.2)$$

where:

- $x^m_j[n]$ is the signal transmitted on the j-th PRB by the m-th MUE;
- $h^m_j(q)$ is the channel gain for the q-th replica of the signal transmitted on the j-th PRB by the m-th MUE; this gain is distributed as a circularly-symmetric complex Gaussian random variable, $h(q) \sim \mathscr{CN}(0, \sigma^2_q)$ according to [12];
- $\tau^m_j(q)$ is the q-th propagation path delay of the signal transmitted on the j-th PRB by the m-th MUE, and is expressed in samples;
- $\theta^m_j(q)$ is the DoA, which represents the angle between the q-th replica of the signal transmitted on the j-th PRB by the m-th MUE and the antenna array axis;
- $v_l \sim \mathscr{N}(0, N_0/2)$ is the AWGN on the l-th antenna.

Similarly, an identical expression can be written for the s-th SUE.

The DoAs of the different clusters of signal replicas that are received by the indoor HeNB vary depending on the environment and the positions of the scatterers. Therefore, it follows that a uniform angular distribution of clusters in the $[0, 2\pi]$ range is a more realistic choice [15, 16].

The propagation delay of the q-th signal between two consecutive antenna elements is $\delta^m_j(q) = d\sin(\theta^m_j(q))/c$. Therefore, when we consider $\tau \ll T_s$, where T_s is the sampling period, the phase of the arriving signal is rotated by $2\pi f_0 \delta^m_j(q)$. By defining the steering vector towards the direction $\theta^m_j(q)$ as $\mathbf{s}(\theta^m_j(q))$, which contains the elements $s_l(\theta^m_j(q)) = e^{j\pi(l-1)\sin(\theta^m_j(q))}$ with $l = 1, 2, \cdots, L$, the m-th MUE signal that is received on the j-th PRB can be expressed as

$$\mathbf{r}^m_j = \mathbf{S}^m_j \cdot \mathrm{diag}(\mathbf{h}^m_j) \cdot \mathbf{x}^m_j + \mathbf{v} \qquad (4.3)$$

where $\mathbf{x}^m_j = [x^m_j[n - \tau^m_j(1)], \cdots, x^m_j[n - \tau^m_j(M)]]^T$, $\mathbf{S}^m_j = [\mathbf{s}(\theta^m_j(1)), \cdots, \mathbf{s}(\theta^m_j(M))]$, $\mathrm{diag}(\cdot)$ is the diagonal matrix, and $\mathbf{h}^m_j = [h^m_j(1), \cdots, h^m_j(M)]$.

The DoA estimation process is performed using an algorithm that belongs to the MuSiC class of algorithms, which are described in detail in [14]. Specifically, the Root-MuSiC algorithm [8], which is a modified version of the most widely known Spectral MuSiC algorithm, has been selected.

MuSiC algorithms are based on eigenvalue decomposition of the received signal autocorrelation matrix of the j-th PRB, $\mathbf{R}^m_j$.

In the first stage, the autocorrelation matrix from the received signal samples is estimated, i.e.,

$$\mathbf{R}_j^m \triangleq E\left[\mathbf{r}_j^m (\mathbf{r}_j^m)^H\right]$$

$$= E\left[\mathbf{S}_j^m \operatorname{diag}(\mathbf{h}_j^m) \, \mathbf{x}_j^m (\mathbf{x}_j^m)^H \operatorname{diag}(\mathbf{h}_j^m)^H (\mathbf{S}_j^m)^H\right] + E\left[\mathbf{vv}^H\right]$$

$$= \mathbf{S}_j^m \mathbf{P}_j^m (\mathbf{S}^m)^H + \sigma_v^2 \mathbf{I}_L$$

where $\mathbf{I}_L$ is the $L \times L$ identity matrix, $[\cdot]^H$ is the Hermitian operator, and $\mathbf{P}_j^m$ is defined as

$$\mathbf{P}_j^m = E\left[\operatorname{diag}(\mathbf{h}_j^m) \, \mathbf{x}_j^m \mathbf{x}_j^{mH} \operatorname{diag}(\mathbf{h}_j^m)^H\right]. \tag{4.4}$$

The eigenvalues of the autocorrelation matrix identify both a signal and a noise subspace [14]. The eigenvectors are thus sorted in descending order based on the value of the corresponding eigenvalue and are then divided into two subspaces. The first subspace is called the signal subspace $\mathbf{U}_S$, and it is composed of the K eigenvectors with the highest eigenvalues. Similarly, the second subspace is called the noise subspace $\mathbf{U}_N$, and is composed of the remaining eigenvectors $(L - K)$ with the smaller eigenvalues. If K is not known, then the two subspaces can be split by relying on the power of the uncorrelated noise incident on the antenna. In fact, all the eigenvalues, λ_i, corresponding to the signal subspace eigenvectors, are greater than the noise power:

$$\lambda_1 > \lambda_2 > \cdots > \lambda_K > \lambda_{K+1} = \cdots = \lambda_L = \sigma_n^2 \tag{4.5}$$

Ideally, the two subspaces are orthogonal to each other because of their algebraic construction, and, they are therefore disjoint. Also, their union contains all the steering vectors. It follows that when we calculate the value of the projection on the noise subspace of each steering vector ($\|\mathbf{s}^H(\theta)\mathbf{U}_N\|$), then the lowest K projection values are those related to the K signal DoAs. The Spectral MuSiC algorithm calculates a function that is inversely proportional to the projection of each steering vector on the noise subspace. When a steering vector is orthogonal to the noise subspace, then its projection is zero, and the function has a local maximum. Specifically, the Spectral MuSiC algorithm searches the K steering vectors $\mathbf{s}(\theta_j^m(k))$ with $k = 1, \cdots, K$ that maximize the function

$$P_{SM}(\theta) = \frac{1}{\|\mathbf{s}^H(\theta)\mathbf{U}_N\|} \tag{4.6}$$

where $\| \cdot \|$ is the vector norm. However, the peaks of (4.6) are strongly influenced by the SNR value received. The Root-MuSiC algorithm attempts to solve this impairment. Localization of the function peaks is replaced by an evaluation of the roots of the polynomial at the denominator of (4.6). This is carried out via a problem reformulation in the z domain, which allows the roots that lie closest to the unit circle to be determined. The denominator of (4.6) for the case of an antenna array with equispaced elements can be written as

$$P_{SM}(\theta)^{-1} = \mathbf{s}^H(\theta)\mathbf{Q}\mathbf{s}(\theta)$$

$$= \sum_{p=1}^{L}\sum_{q=1}^{L} e^{-j\pi(p-1)\sin\theta} Q_{p,q} e^{j\pi(q-1)\sin\theta}$$

$$= \sum_{l=-L+1}^{L-1} q_l e^{-j\pi l \sin\theta}$$

where $\mathbf{Q} = \mathbf{U}_N\mathbf{U}_N{}^H$ and $q_l \triangleq \sum_{l=-q} Q_{p,q}$ is the sum of the elements of $\mathbf{Q}$ on the l-th diagonal. Evaluation of the peaks of $P_{SM}(\theta)$ is equivalent to computation of the polynomial roots of $D(z)$ on a unit circle, where the polynomial is defined as

$$D(z) = \sum_{l=-L-1}^{L-1} q_l z^{-l} \qquad (4.7)$$

This leads to improved DoA estimation accuracy for low SNR values, and allows even close signals to be distinguished. In addition, the Root MuSiC algorithm does not require the selection of a suitable threshold to detect the maximum values of the function.

However, when the number of received signals K is higher than $L-1$, the splitting of the two subspaces into $L-1$ and 1 is enforced. Consequently, the noise subspace is subject to interference by the useful signal and, these quantities are no longer disjoint. This leads to a loss of the DoA estimation accuracy that generally increases with K.

In the scenario considered here, the number of received signals, K, is given by the number of UEs in the HeNB coverage area, $(F+N)$, multiplied by the number of multipath components, M. However, by operating on the PRB basis, it is possible to separate the user signals, because each PRB is allocated to a different user. As a result, the Root MuSiC algorithm is performed separately for each PRB, allowing for estimation of up to $(L-1)*J$ DoAs. When the number of incident signals per PRB is more than $L-1$, then only the most powerful signals are selected.

In summary, the DoA estimation algorithm is able to estimate up to $L-1$ DoAs per PRB, and thus can estimate up to $L-1$ DoAs per UE. Consequently, the term K is defined as

$$K = \begin{cases} M & \text{if } M \leq L\text{-}1 \\ L-1 & \text{otherwise} \end{cases} \qquad (4.8)$$

This means that:

- If $M < L-1$, then the M estimated DoAs correspond to all received signals;
- If $M > L-1$, then only the most powerful $L-1$ estimated DoAs are detected.

The DoA estimation is updated every subframe to take time variations because of UE mobility into account.

The effects of $M > L - 1$ on both beamforming and DoA estimation have been considered in [1], and are reported in Sect. 4.3.3.

4.3.2 Zero-Forcing Beamforming

The ZFBF methodology allows the HeNB to place nulls in the radiation pattern in the directions of the MUEs placed in its coverage area by using suitable transmission weights. The estimated DoAs are therefore used as inputs to a ZFBF algorithm that returns L complex numbers; these numbers are used to adjust the amplitude and phase of the output signal from each antenna by multiplication of the time domain signal.

The algorithm limits are inherited from linear array theory. Therefore, each weight vector is able to model $L - 1$ nulls. This is insufficient to reduce the interference of all the MUEs that can be found within the HeNB coverage area. For this reason, as per the DoA estimation algorithm, null steering is performed for the PRB. One weight vector is generated for each PRB and is then used to weight the subcarriers that belong to that PRB. By operating in this way, different weights and nulls are obtained in different positions for each group of 12 subcarriers.

Assuming that the j-th PRB is allocated to the m-th MUE by the eNB and to the f-th SUE by the HeNB, the weights calculated by the HeNB are then obtained by imposing a projection on $\mathbf{s}(\theta_j^f(1))$ equal to 1 and a projection on $\mathbf{s}(\theta_j^m(1)), \cdots, \mathbf{s}(\theta_j^m(K))$ equal to 0 [8]. Using the notation where

- $\mathbf{A_{j,f}} = \left[\mathbf{s}(\theta_j^f(1)), \quad \mathbf{s}(\theta_j^m(1)), \quad \mathbf{s}(\theta_j^m(1)), \quad \cdots, \mathbf{s}(\theta_j^m(K))\right]$ is a matrix containing the steering vectors of interest for the j-th PRB (which has dimensions of $L \times (K + 1)$);
- $\mathbf{c} = [1, 0, \cdots, 0]^T$ is a vector with $K + 1$ elements,

then the resulting beamforming weights are

$$\mathbf{w}_{j,f}^H = \mathbf{c}^T \mathbf{A}_{j,f}^H (\mathbf{A}_{j,f} \mathbf{A}_{j,f}^H)^{-1}. \tag{4.9}$$

where $\mathbf{w}_{j,f}$ is a vector of length L where the element $w_{j,f}(l)$ is used to weight the signal transmitted on the j-th PRB (which is assigned to the f-th SUE) by the l-th antenna.

4.3.3 Performance Evaluation

This section presents some numerical results to demonstrate the ability of the interference mitigation scheme to effectively reduce the interference directed towards the MUEs when the algorithm works on a single PRB.

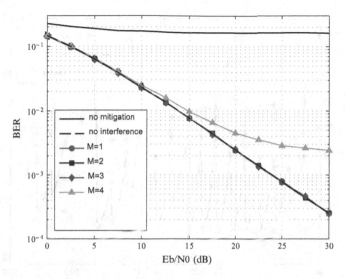

Fig. 4.2 Macrocell DL BER performance when ideal DoA estimates are available at the HeNB [1]

To aid with the description of the results, the following notations are used:

- $link_1$ is the link between the MUE and the HeNB used for DoA estimation;
- $link_2$ is the macrocell DL that connects the eNB and the MUE.

First, the ideal DoA estimation case is considered. Figure 4.2 shows the BER performance experienced by the MUEs in DL communications, when the previously described approach to mitigate interference is applied by assuming perfect knowledge of the DoAs for the received signals on the HeNB side.

The BER performance is given as a function of the ratio between the received energy per bit, E_b, and the single-sided noise spectral density, N_0, of $link_2$ for different numbers of resolvable multipath components. Note here that if the multipath components that are introduced by the propagation channel are fewer than L in number, then the system is able to eliminate the small cell interference completely. The performance degrades when the number of resolvable paths is higher than $L - 1$ (i.e., $M = 1$) and there is therefore residual interference because of the fourth path, which cannot be deleted.

Obviously, DoA estimation errors reduce the effectiveness of the interference mitigation scheme, because ZFBF then places the nulls in the wrong directions. The DoA estimation accuracy is thus of paramount importance, and this accuracy is analyzed in the following, assuming actual propagation conditions that mean that the UEs have mobility and the number of resolvable paths is higher than $L - 1$.

In the mobility context, the channel impulse response and the signal DoA vary with time, depending on the speed of the UE. The DoA estimation algorithm must therefore be able to accommodate these variations. Figure 4.3 shows the SIR value at the MUE receiver, which was determined using an interference mitigation

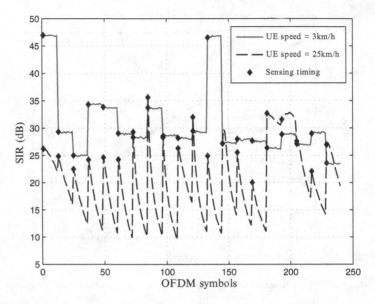

Fig. 4.3 Received SIR on $link_2$ at the MUE side with DoA estimation and ZFBF (where $M = 3$) [1]

scheme based on ZFBF under various MUE mobility speed conditions. The results have been derived by assuming that, without interference mitigation, the MUE experiences a SIR value of 0 dB and that the DoA estimation is updated for every subframe. Assuming pedestrian-level mobility (3 km/h), the tracking algorithm can follow the DoA variations between two successive sensing intervals. Therefore, the interference power level variations are limited because of the imperfect positioning of the null beams. As the UE speed increases, there is a corresponding increase in the interference between the two sensing intervals because of the rapid channel variability. This results in degradation of the SIR value between the two sensing intervals. However, a significant improvement in SIR is obtained even in this case when compared with the case of the absence of interference mitigation. These results are also clearly shown in Fig. 4.4, where the performance is shown in terms of BER for various UE speeds as a function of E_b/N_0 for $link_2$. It is shown that there is only a slight increase in BER with speed for high E_b/N_0 values.

Figure 4.5 presents the DoA estimation error of the Root-MuSiC algorithm as a function of E_b/N_0 received on $link_1$. The value of the *weighted Euclidean DoA error* is computed by weighting the error with weights g_i, where $i = 1, \cdots, M$, that are proportional to the magnitude of each replica.

The DoA estimation accuracy deteriorates when the number of resolvable paths increases. Up to $M = 3$, all resolvable paths can be detected, but when the number of signal replicas increases, the noise on the weaker replicas leads to an increase in the estimation error. When $M = 4$, the intrinsic limitation of the DoA estimation algorithm is overcome (i.e., more than $L - 1 = 3$ replicas are received) and the signal and noise subspaces are not more disjointed.

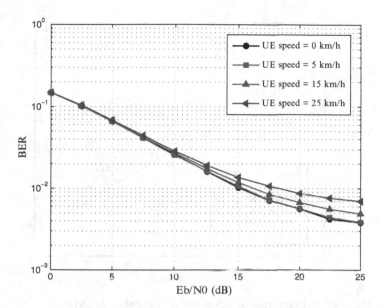

Fig. 4.4 Macrocell DL BER performance for various MUE speeds ($M = 3$, $SIR = 0$ dB on $link_2$) [1]

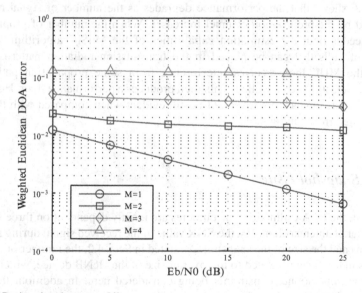

Fig. 4.5 DoA estimation errors [4]

However, even when the number of signal replicas is higher than $L - 1$, the interference mitigation scheme analyzed here leads to a significant improvement in the macrocell DL performance, as shown in Fig. 4.6. The BER values are derived as a function of the $link_2$ E_b/N_0 when the $link_1$ E_b/N_0 value is 10 dB.

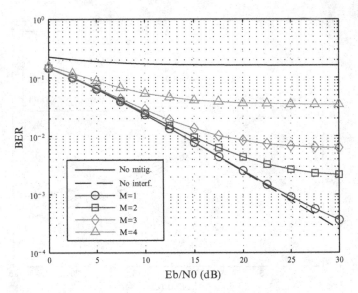

Fig. 4.6 Macrocell DL BER performance, where E_b/N_0 of $link_1 = 10\,\mathrm{dB}$ [4]

Figure 4.6 shows that the performance degrades as the number of signal replicas generated by the propagation channel increases. When the number of propagation paths exceeds both the estimation capabilities of the MuSiC algorithm and the number of nulls that can be placed in the transmission radiation pattern by the HeNB, the MUE then receives interference from some propagation paths that cannot be detected and cancelled. This then produces a performance floor. However, significant performance improvement is still evident when compared with the case without interference mitigation.

4.3.4   Snapshot Issue

In general, the DoA estimation inaccuracies are mainly dependent on three factors: the number of antenna elements, the SNR value of the received signal during sensing operations and the size of the snapshot. As stated in Sect. 4.2, the number of antenna elements in the array is related to the overall size of the HeNB device, which is one of the main operational constraints being considered here. In addition, the DoA estimation algorithm selected can provide high accuracy, even when the received signal is degraded (i.e., low SNR values). The third factor, the snapshot size, has been widely addressed in the literature [8, 11]. MuSiC algorithms usually work optimally for oversized snapshots that are not available at the receiver end in certain contexts. The main solution that has been proposed in the literature is reliance on high numbers of antennas, ranging up to double the number of signals that are to be detected. Alternatively, the data received during the various sensing stages could be temporarily collected for averaging of the values of the autocorrelation matrix, thus reducing the effects of noise and leading to improved estimates. However, the MUE resource scheduling should remain unchanged for the averaging time, but this cannot be guaranteed. The best solution could therefore be to increase the snapshot dimension in the frequency domain by gathering together the PRBs that belong to the same user. This would also lead to reduced computational complexity, because the Root-MuSiC and ZFBF algorithms would have to be run fewer times.[3] However, this requires knowledge of the UL resource allocation for the MUEs. In certain scenarios, this knowledge can be achieved using a limited signaling exchange between the cells [7, 9, 13]; otherwise, this problem can be addressed by introducing algorithms that can determine which PRBs belong to the same user. Two possible methods to achieve this aim are:

- Preliminary rough estimation of the DoA of the main signal component on each PRB;
- Use of the similarity of the signal eigenspaces belonging to the same user [2].

The second method in particular, which originated in [2], is based on projection of the most informative eigenvector of each PRB on the space represented by the main eigenvectors of the other PRBs. The same UE in fact has similar eigenvectors in the signal eigenspace of each of the PRBs allocated to it. Therefore, correlation between the eigenvectors of PRBs of the same UE is high, and this makes it possible to identify the groups of PRBs that belong to each user.

As stated previously, the DoA estimation capabilities can be dependent on the snapshot dimension used by the Root-MuSiC algorithm and, consequently, on the number of PRBs used per UE. The worst case scenario occurs when PRB clustering is not available (i.e., it is not possible to estimate which PRBs belong to the same user) or each PRB is allocated to a different user. It therefore follows that the size of the snapshot used for DoA estimation is equal to the PRB size, and some inaccuracies are introduced when the number of incident signals is high. The upper bound of the estimation performance is reached when all PRBs are assigned to a single user, and this information is available at the HeNB. In this case, the snapshot is composed of all the subcarriers, and the system discussed in this section also achieves good performance for numbers of incident signals that are higher than $L - 1$. This specific case coincides with consideration of an OFDM system, such as that in [1].

Figure 4.7 shows the accuracy of the DoA estimates as a function of the number of PRBs used to compose the snapshot for sensing when $M = 4$. For each abscissa value, there are three bars that represent the DoA estimation error occurrences of the three $(L - 1)$ estimated paths. Different shades of gray are used to distinguish the different error intervals: $[0, 2.5°], [2.5°, 5°], [5°, 10°]$. It is shown that the first path is always detected with high accuracy, even when the number of PRBs is limited.

---

[3]They would be run for each group of PRBs rather than for each PRB.

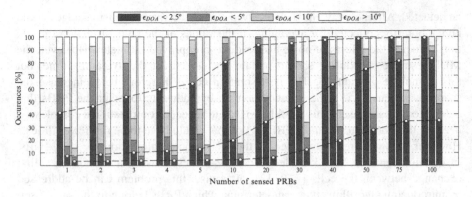

Fig. 4.7 DoA estimation precision as a function of PRBs used for sensing [4]

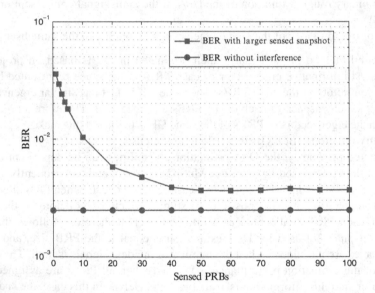

Fig. 4.8 Macrocell DL BER performance as a function of the number of PRBs used for sensing [4]. $M = 4$, $E_b/N_0 = 10\,\mathrm{dB}$ on $link_1$, $E_b/N_0 = 20\,\mathrm{dB}$ on $link_2$

Conversely, the second and third path DoA estimates show good precision with increasing numbers of sensed PRBs.

Clearly, by gathering the PRBs that belong to the same UE together, it is possible to improve the performance of the described interference suppression method significantly. This is shown in Fig. 4.8. The BER of the interference mitigation system is reported as a function of the PRBs used to compose the snapshot when $M = 4$. The BER value results achieved are very close to the results for the case without interference. Finally, it is important to note that significant performance

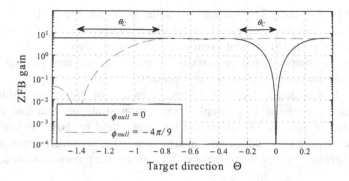

Fig. 4.9 ZFBF gain vs. target direction for various null direction values [5]

improvements can be achieved by increasing the number of PRBs up to 20, where the saturation point is then reached.

4.4   Cognitive Resource Allocation with Spatial User Selection

The interference mitigation scheme discussed in Sect. 4.3.3 reduces the interference directed towards the MUEs, but it does not take the requirements of the SUEs into account. However, the beamforming gain depends not only on the null direction, ϕ_{null}, but also on the SUE direction, Θ, which is simultaneously served using the same resources. This represents the *target direction* in which the maximum beamforming gain is desired. The ZFBF scheme can guarantee almost zero interference in the concurrent MUE direction and near-maximum gain in the target SUE direction if the difference between the two DoAs is sufficiently high, as shown in Fig. 4.9, where the ZFBF gain is provided for two values of ϕ_{null} as a function of the *target direction* Θ. From this figure, it is evident that the selection of the pair of users served by the same PRB affects the system performance.

For this reason, the previously described interference reduction scheme can be combined with a suitable selection of macrocell and small cell UEs that can be served on the same PRB. Two specific cognitive resource allocation methods are investigated, based on new user selection criteria, with the goal of maximizing the small cell capacity and counteracting the interference directed towards the macrocell [3]. Both methods operate by using their knowledge of the signal DoAs. The quality of the methods considered here is demonstrated by providing performance comparisons with other benchmark methods in terms of both small cell capacity and the BER performance at the macrocell DL.

The method illustrated here focuses specifically on a heterogeneous cellular system, where each PRB is simultaneously assigned to both a macrocell and a small cell user, which are selected using procedures based on knowledge of the signal DoAs. For each RU, the optimal solution selects the SUE that achieves the maximum beamforming gain by imposing the condition that the interference directed towards the MUE that communicates in that resource block is zeroed. To reduce the computational complexity, a faster heuristic method that is based on the DoA information alone is also presented. The most important property of this method is that it does not require evaluation of the beamforming gains. Both methods represent very simple transmission strategies that can easily be implemented in practice. An important advantage held by both methods is that they do not require knowledge of the interference link CSI, which is unavailable in cognitive networks, where the primary system ignores the presence of the secondary system.

Finally, it should be noted that there is no coordination between the cells, but it is assumed that only the scheduling information for the MUEs (both for the UL and the DL) is provided by the eNB through the backhaul or an air interface [7, 9, 13]. This information is used together with the sensing results to associate the DoA estimates for the MUE signals with each DL PRB.

The key features of this framework are:

- The HeNB receives the scheduling information for the MUEs for both UL and DL from the eNB;
- The HeNB performs spectrum sensing during UL transmission to estimate the DoAs of the signals received by the MUEs and SUEs, as described in Sect. 4.3.3;
- The HeNB compares the spectrum sensing results with the scheduling information to associate the DoA estimates with the allocated DL resources;
- The HeNB allocates resources to its SUEs by following a suitable policy that allows it to null the interference towards the MUEs and maximize the small cell capacity.

For simplicity, the condition $K = M$ is assumed in the following discussion.

4.4.1 Detailed Procedure

Let us assume that the eNB has independently assigned the j-th PRB to the $\bar{m}$-th MUE for DL communications, and that this MUE is in the coverage area of the small cell. During the sensing phase, the HeNB estimates the DoA of the signals transmitted by the $\bar{m}$-th MUE, and obtains $\theta_j^{\bar{m}}(1), \theta_j^{\bar{m}}(2), \cdots, \theta_j^{\bar{m}}(M)$. Then, the HeNB must *select* the SUE that enables avoidance of interference being directed towards the $\bar{m}$-th MUE while simultaneously maximizing the cell capacity. Two particular methods are illustrated here. The first is an optimum method that selects the SUE that achieves the MBG, while the second is a sub-optimal method, called the location aware (LA) method, which selects the SUE that presents the maximum angular distance to the $\bar{m}$-th MUE. In both cases, the goal of the resource allocation problem is to determine the allocation matrix Ω, which has dimensions of $F \times J$ and is defined as

$$\Omega[f,j] = \begin{cases} 1 & \text{if } j-th \text{ PRB is assigned to user } f \\ 0 & \text{otherwise} \end{cases} \tag{4.10}$$

At the beginning of the procedure, all elements of the Ω matrix are set to zero. Then, iteratively for each PRB, the procedure selects the $\hat{f}$-th SUE, which is defined as

$$\hat{f} = \arg\max_{f=1,\cdots,F} \{O_j(i)\}, \qquad j = 1,\cdots,J. \tag{4.11}$$

Depending on the method selected, two different objective functions, $O_j^{MBG}(f)$ and $O_j^{LA}(f)$, are explained in the following. Also, if the j-th PRB is allocated to user $\hat{f}$, then element $(\hat{f},j)$ of matrix Ω is set to 1: $\omega[\hat{f},j] = 1$.

4.4.1.1   User Selection: MBG

For each PRB_j, where $j = 1,\cdots,J$, and for each SUE_f, where $f = 1,\cdots,F$, the algorithm evaluates the matrix $A_{j,f}$ that contains the steering vectors and the beamforming weights, as in (4.9). On that basis, the objective function is equal to the beamforming gain of the f-th SUE directed towards its main DoA, which can be expressed as

$$O_j^{MBG}(f) = G(\theta_j^f(1)) = |w_{j,f}^H s(\theta_j^f(1))|^2 = |s^H(\theta_j^f(1))(A_{j,f}A_{j,f}^H)^{-1}s(\theta_j^f)|^2. \tag{4.12}$$

4.4.1.2   User Selection: LA

The LA user selection criterion is a sub-optimal method that can be used to solve the MBG problem with reduced complexity. This criterion is based on the idea that, in general, the beamforming gain increases by the same extent that the SUE to MUE angle separation increases. Thus, for each PRB, the criterion selects the SUE that presents the highest angular distance to the MUE that was allocated in the PRB. This scheme does not require any additional computation other than the DoA estimates. To describe the scheme in more detail, for each PRB j, where $j = 1,\cdots,J$ and each SUE f, where $f = 1,\cdots,F$, the algorithm calculates the angular distances between the $\bar{m}$-th MUE using the j-th PRB and all the SUEs as

$$d_{f,j} = \left| \theta_j^f(1) - \theta_j^{\bar{m}}(1) \right| \tag{4.13}$$

and thus the resulting objective function is

$$O_j^{LA}(i) = d_{f,j}. \tag{4.14}$$

As a final remark, it should be noted that the analysis and simulation results presented below consider the worst case scenario: where all the PRBs are used by MUEs in the small cell coverage area. In general, however, only a few of the PRBs are used by the MUEs in the small cell areas and the sensing phase can detect whether the PRBs are available or not (e.g., by using an energy detection strategy, as discussed in Sect. 2.4.2). When a PRB is not used by any MUE in the small cell area, the HeNB can then perform conventional maximum gain beamforming and user selection depends solely on the scheduling policy.

4.4.2 Performance Evaluation

This section presents numerical results that were obtained using computer simulations [3]. With the specific goal of demonstrating the quality of the previously described schemes, the performance of these schemes is compared in terms of both capacity and BER with that of

- A conventional beamforming (conventional-BF) scheme that maximizes the small cell capacity without considering the presence of the MUE. The beamforming gain in the direction of the main propagation path (i.e., the *target direction*) is always $G(\theta_i^F(1)) = L$, and the user allocation is random. This represents the upper bound for small cell performance and the lower bound for macrocell performance because the interference is not managed;
- A ZFBF scheme, where the SUE resource assignment is random (ZFBF-random allocation). This represents the lower bound for small cell performance and the upper bound for macrocell performance because it focuses solely on interference nulling (this is the method that is described in Sect. 4.3).

In conventional BF and ZFBF schemes, where user selection is not performed, the allocation follows a RR scheduling approach.

In deriving the results, the following working hypotheses are assumed:

- Bandwidth $B = 2.5$ MHz
- Number of PRBs $J = 12$
- Multipath indoor propagation channel in accordance with that of [12].
- Number of SUEs $F = 3$
- Number of antenna elements $L = 4$.

Figure 4.10 shows the mean capacity of the small cell as a function of E_b/N_0. As expected, the highest capacity is obtained with conventional BF, while ZFBF with random allocation represents the worst case scenario. The methods illustrated in Sect. 4.4.1 have produced results that are close to the upper bound by significantly improving the random allocation technique. In addition, it should be noted that the sub-optimal LA resource allocation scheme produces a performance that is very close to that of the optimal MBG, but with the important advantage of reduced computational complexity.

Fig. 4.10 Small cell capacity using user selection policies [3]

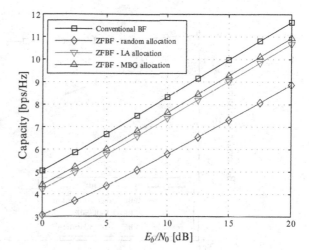

Fig. 4.11 BER of the small cell DL [3]

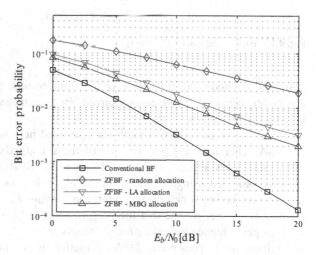

The benefits of the resource allocation methods that are illustrated here must also be evaluated by taking the BERs of the SUEs and MUEs into account. The goal of these schemes is to obtain a good trade-off between increased small cell capacity and reduced interference directed toward the MUEs. Figures 4.11 and 4.12 show the DL BERs of a SUE and a MUE in the small cell area, respectively. The curves are derived as a function of E_b/N_0 of the related links (i.e., E_b/N_0 of the small cell DL in Fig. 4.11, and E_b/N_0 of the macrocell DL, $link_1$, in Fig. 4.12), using uncoded QPSK modulation.

The behavior of the curves in Fig. 4.11 follows that of the capacity in Fig. 4.10, and the same conclusions can be drawn. With regard to the MUEs (Fig. 4.12), the

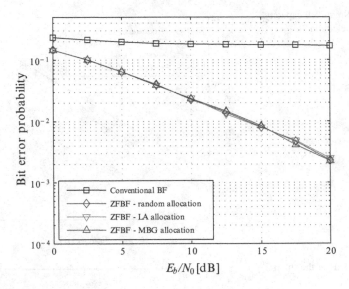

Fig. 4.12 BER of the macrocell DL [3]

described methods and ZFBF achieve the same results, because the interference is completely removed in both cases. Conversely, the conventional BF method presents very high BER values.

The interference reduction methods discussed here are based on knowledge of the DoA. Therefore, the performance of these methods can be affected by DoA estimation errors. Figures 4.13 and 4.14 show the BERs of the two cells, when the DoA is estimated using the Root-MuSiC algorithm, as a function of the sensing E_b/N_0. For the small cell in particular, the sensing E_b/N_0 corresponds to the DL E_b/N_0. Therefore, we see that the performance follows the ideal DoA case with obvious performance degradation that mainly affects the conventional BF scheme. In addition, in the presence of DoA estimation errors, the differences between the two proposed methods, MBG and LA, tend to be reduced. The DoA estimation errors therefore mainly affect the determination of the maximum BF gain. In contrast, in Fig. 4.14, the sensing E_b/N_0 is that of $link_2$, while the E_b/N_0 of the DL MUE is fixed at 12 dB. It can be seen that the performance only shows a little improvement when the sensing E_b/N_0 increases, because the Root-MuSiC algorithm can even provide good estimates for low values. The BER degradation achieved by the MUE is comparable with that of the small cell.

Fig. 4.13 BER of the small
cell DL with DoA estimation
errors [3]

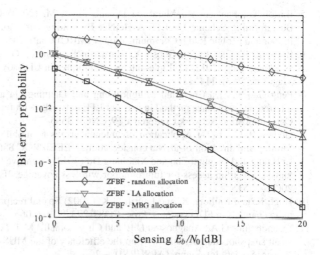

Fig. 4.14 BER of the
macrocell DL with DoA
estimation errors [3]

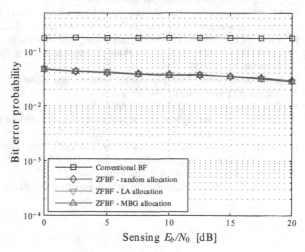

References

1. Bartoli G., Fantacci R., Marabissi D. and Pucci M. (2013) LTE-A Femto-Cell Interference Mitigation with MuSiC DOA Estimation and Null Steering in an Actual Indoor Environment. Proc. IEEE International Conference on Communications (ICC): 1–5.
2. Bartoli G., Fantacci R., Marabissi D. and Pucci M. (2014) Physical Resource Block clustering method for an OFDMA cognitive femtocell system. Physical Communication Elsevier 11.
3. Bartoli G., Fantacci R., Marabissi D. and Pucci M. (2014) Resource allocation schemes for cognitive LTE-A femto-cells using zero forcing beamforming and users selection. IEEE Global Communications Conference (GLOBECOM): 3447–3452
4. Bartoli G., Fantacci R., Marabissi D. and Pucci M. (2014) Angular interference suppression in cognitive LTE-A femtocells. Wireless Communications and Mobile Computing Conference (IWCMC): 979–984

5. Bartoli G., Fantacci R., Marabissi D. and Pucci M. (2014) Coordinated Scheduling and Beamforming Scheme for LTE-A HetNet Exploiting Direction of Arrival. IEEE Personal, Indoor and Mobile Radio Communication (PIMRC) Conference
6. Boccardi F., Clerckx B., Ghosh A., Hardouin E., Jongren G., Kusume K., Onggosanusi E. and Yang Tang (2012) Multiple-antenna techniques in LTE-advanced. IEEE Communication Magazine 50(3): 114–121
7. Peng Gao and Da Chen and Mingjie Feng and Daiming Qu and Jiang Tao (2013) On the interference avoidance method in two-tier LTE networks with femtocells. Proc. IEEE Wireless Communication Network Conference (WCNC): 3585–3590
8. Godara, L.C. (1997) Application of antenna arrays to mobile communications. II. Beamforming and direction-of-arrival considerations. IEEEJPROC 85(8):1195–1245
9. Li Huang and Guangxi Zhu and Xiaojiang Du (2013) Cognitive femtocell networks: an opportunistic spectrum access for future indoor wireless coverage. IEEE Wireless Communication 20(2):44–51
10. Hugl, K. and Kalliola, K. and Laurila, J.K. (2002) Spatial reciprocity of uplink and downlink radio channels in FDD systems. Proc. COST 273 TD(02) 066
11. Ioannopoulos G.A., Anagnostou D.E. and Chryssomallis M.T. (2012)A survey on the effect of small snapshots number and SNR on the efficiency of the MUSIC algorithm. IEEE Antennas Propag. Society Int. Symp. (APSURSI):1–2.
12. ITU-R International Telecommunication Union Recommendation (1997) Guidelines for evaluation of radio transmission technologies for IMT-2000. M.1225
13. Sahin M.E., Guvenc I., Moo-Ryong Jeong and Arslan, H. (2009) Handling CCI and ICI in OFDMA femtocell networks through frequency scheduling. IEEE Transaction Consum. Electron. 55(4):1936–1944
14. Schmidt R. (1986) Multiple emitter location and signal parameter estimation. IEEE Trans. Antennas Propag. 34(3):276–280
15. Q.H. Spencer and B.D. Jeffs and M.A. Jensen and A.L. Swindlehurst (2000) Modeling the statistical time and angle of arrival characteristics of an indoor multipath channel. IEEE Journal of Selected Area on communications 18(3):347–360
16. Lei Jiang and Soon Yim Tan (2007) Geometrically Based Statistical Channel Models for Outdoor and Indoor Propagation Environments. IEEE Trans. Veh. Technology 56(6): 3587–3593
17. 3GPP – Third Generation Partnership Project (2012) Evolved Universal Terrestrial Radio Access (E-UTRA) – Base Station (BS) radio transmission and reception. TS36.104

Chapter 5
Conclusions

5.1 Brief Recap

The demand for pervasive wireless access and requirements for high data rates are expected to grow significantly in the near future. This process will be triggered by the huge proliferation of high resource demand applications such as gaming, mobile television and specific wireless services. In this context, the deployment of HetNets will enable important capabilities such as high data rates and traffic offloading to be realized, and will provide dedicated capacity to homes, enterprises, and urban hotspots. HetNets represent a novel networking paradigm based on the concept of deploying short-range, low-power, and low-cost base stations that operate in conjunction with the main 5G macrocellular network infrastructure. HetNets encompass a broad variety of cells, including microcells, picocells, metro cells, and femtocells, along with advanced wireless relays and distributed antennas that can be deployed anywhere.

However, mass deployment of HetNets introduces certain challenges that must be addressed to ensure that the expected benefits materialize. Among these challenges, interference management is one of the most important considerations in co-channel HetNet deployment, where the small cells and the macrocells share the same frequency spectrum. Different approaches can be adopted that depend on aspects such as the specific scenario, the computational capacity of the devices, and the available backhaul link.

This Brief attempts to provide an overview of the inter-layer interference problems of HetNets and provide a critical outline of the possible solutions for practical implementation.

After the introduction of the main concepts and possible deployment scenarios of HetNets, Chap. 1 outlined the emerging research challenges. Chapter 2 then explained the main concepts of interference management in HetNets when deployed using a spectrum sharing approach, and provided an overview of the best-known

© The Author(s) 2015   73
D. Marabissi, R. Fantacci, *Cognitive Interference Management in Heterogeneous Networks*, SpringerBriefs in Electrical and Computer Engineering,
DOI 10.1007/978-3-319-20191-7_5

techniques. In the second part of this Brief, the focus has been placed on cognitive HetNet technology from the perspective of the application of this technology to 5G networks. In 4G networks, interference problems are solved by resorting to coordinated and cooperative approaches. However, this methodology is only viable when both the macrocells and the small cells are directly deployed by the same operator; otherwise, the coexistence and efficient coordination of the macrocells and small cells becomes very challenging. This drawback is accentuated in 5G networks, where the number of small cell nodes is expected to increase significantly, and many user-deployed small cells will be used in various environments, including homes, small offices and enterprises. In such a scenario, coordination between the two network layers to manage inter-cell interference will either be infeasible or impossible, mainly because of network delays and signaling overheads. This problem may be addressed by the emerging paradigm of cognitive HetNets, in which each small cell is provided with sensing capabilities to acquire knowledge about the macrocell transmissions and then adapt its own transmission/reception behavior using opportunistic and dynamic resource allocation schemes and advanced signal processing methods. Consequently, Chap. 3 provided a description of the most popular cognitive methodologies intended for use in cognitive HetNets and highlighted their advantages and challenges for practical implementation. To better understand the cognitive HetNets paradigm and to provide the reader with knowledge of the practical tools and methodologies, Chap. 4 discussed one possible cognitive approach that was able to reduce co-channel interference while simultaneously increasing the small cell capacity on the basis of a multiple-antenna technology approach. The goal of this method is to use the spatial dimension as a new spectrum opportunity to enable the coexistence of small cells with macrocells. The small cell models its transmissions by placing nulls in the directions of the MUEs, thus protecting the macrocell DL transmissions. Also, the small cell performs an appropriate resource allocation process for its UEs to increase its communication capacity.

5.2 Future Challenges

Cognitive small cells offer a particularly promising solution for indoor home- and enterprise-based communications and applications. These cells will be an important instrument in enabling UE demands to be met by allowing high access network flexibility and reconfigurability. However, many challenges will have to be investigated and overcome for successful operation of cognitive small cells in future 5G networks.

First, the main characteristics of cognitive small cells are their *cognitive* and *self-configuration* capabilities. Cognitive capabilities refer to the ability of the small cells to acquire reliable awareness of their surrounding environment, and awareness of the spectrum holes of the macrocell in particular. In this context, there are many issues that must be taken into account, including the rapidity of resource assignment when compared with the sensing delay, the channel reciprocity, the availability of limited information exchange with the macrocell, and the ability to identify not only the idle resources, but also any interference-free RUs that are allocated far away from the small cells. The sensing information is then used by the self-configuration feature, which tries to optimize parameters in all layers to improve the performance (e.g., in terms of capacity or energy savings) by reducing the interference directed towards the primary system via opportunistic interference avoidance and cancellation techniques. However, because of the expected massive deployment of small cells, the full advantages of cognitive small cells will only be achieved by adding *self-organizing* features and creating SO-HetNets. This will require increased self-organization capabilities to tune large numbers of parameters that are extended to most parts of the network (e.g., planning, deployment, and operation). At the same time, there is also a need to develop simple SO algorithms that can be implemented on devices with limited power consumption and computing capabilities.

Another important topic is that of MIMO systems. The use of *multiple antenna systems* will be improved via 3D-MIMO techniques, enabling control of beam-forming in both the horizontal and vertical directions. Also, better performance levels will be achieved with larger numbers of antennas when using massive MIMO antenna technology. This allows very narrow beamforming, thus compensating for propagation losses and making it possible to follow the motion of individual UEs. The prospective use of *position-related information* may be very attractive if it can be integrated with mobility management schemes and cell association procedures to provide more balanced loading among the cells. Finally, in terms of *power management*, a cognitive power sleep mechanism can be adopted that uses traffic prediction to shut down the small cells when they are not required, thus reducing interference, costs and power consumption.

Glossary

Beamforming   Digital elaboration of the transmitted and received signal to model suitable radiation patterns.

Cognitive Radio   A radio transmitter/receiver able to detect which portion of the spectrum is currently in use and to jump into the temporarily unused spectrum.

Cognitive Network   It is a network that allows the presence of a primary and a secondary cognitive system sharing the resources in an opportunistic way. The cognitive secondary system must be able to share the spectrum without causing harmful interference.

CoMP   A simultaneous and coordinated transmission of different TPs on the same resources.

CS/CB   It is a CoMP technique where different TPs transmit simultaneously towards different UEs using suitable beamforming techniques and resource allocation strategies.

eICIC   A macrocell and small cell joint resource allocation strategy based on a time division access.

HetNet   A network composed by overlapping heterogeneous cells of different dimensions and capabilities.

JT   It is a CoMP techniques where different TPs transmit simultaneously towards the same UE.

MIMO   Multiple antenna systems used to transmit and receive signals.

Spectrum Sensing   Listening of the radio environment to detect the activity of other radio systems in the area.

Zero Forcing Beamforming   A beamforming technique that places nulls in specific directions.

© The Author(s) 2015
D. Marabissi, R. Fantacci, *Cognitive Interference Management in Heterogeneous Networks*, SpringerBriefs in Electrical and Computer Engineering,
DOI 10.1007/978-3-319-20191-7

Printed in the United States
By Bookmasters